T0181307

RaumFragen: Stadt – Region – Landschaft

RaumFragen: Stadt—Region—Landschaft | SpaceAffairs: City—Region—Landscape

Im Zuge des „spatial turns“ der Sozial- und Geisteswissenschaften hat sich die Zahl der wissenschaftlichen Forschungen in diesem Bereich deutlich erhöht. Mit der Reihe „Raum-Fragen: Stadt—Region—Landschaft“ wird Wissenschaftlerinnen und Wissenschaftlern ein Forum angeboten, innovative Ansätze der Anthropogeographie und sozialwissenschaftlichen Raumforschung zu präsentieren. Die Reihe orientiert sich an grundsätzlichen Fragen des gesellschaftlichen Raumverständnisses. Dabei ist es das Ziel, unterschiedliche Theorieansätze der anthropogeographischen und sozialwissenschaftlichen Stadt- und Regionalforschung zu integrieren. Räumliche Bezüge sollen dabei insbesondere auf mikro- und mesoskaliger Ebene liegen. Die Reihe umfasst theoretische sowie theoriegeleitete empirische Arbeiten. Dazu gehören Monographien und Sammelbände, aber auch Einführungen in Teilaspekte der stadt- und regionalbezogenen geographischen und sozialwissenschaftlichen Forschung. Ergänzend werden auch Tagungsbände und Qualifikationsarbeiten (Dissertationen, Habilitationsschriften) publiziert.

In the course of the “spatial turn” of the social sciences and humanities, the number of scientific researches in this field has increased significantly. With the series “Raum-Fragen: Stadt—Region—Landschaft” scientists are offered a forum to present innovative approaches in anthropogeography and social space research. The series focuses on fundamental questions of the social understanding of space. The aim is to integrate different theoretical approaches of anthropogeographical and social-scientific urban and regional research. Spatial references should be on a micro- and mesoscale level in particular. The series comprises theoretical and theory-based empirical work. These include monographs and anthologies, but also introductions to some aspects of urban and regional geographical and social science research. In addition, conference proceedings and qualification papers (dissertations, postdoctoral theses) are also published.

Olaf Kühne · Lara Koegst

Land Loss in Louisiana

A Neopragmatic Redescription

Olaf Kühne
Universität Tübingen
Tübingen, Germany

Lara Koegst
Universität Tübingen
Tübingen, Germany

The contents of this publication are based on results that have been gathered within the framework of the project "Geographies of Unsustainability – a Neopragmatic Regional Geography of Louisiana" funded by the German Research Foundation (DFG). The authors would like to thank the DFG for making this research possible.

ISSN 2625-6991 ISSN 2625-7009 (electronic)
RaumFragen: Stadt – Region – Landschaft
ISBN 978-3-658-39888-0 ISBN 978-3-658-39889-7 (eBook)
https://doi.org/10.1007/978-3-658-39889-7

This Springer VS imprint is published by the registered company Springer Fachmedien Wiesbaden GmbH, part of Springer Nature.
The registered company address is: Abraham-Lincoln-Str. 46, 65189 Wiesbaden, Germany

Synopsis

This book is oriented on testing and developing the neopragmatic approach of horizontal geographies, in which we follow approaches from the natural sciences, the social sciences, and cultural studies. This is done by means of a rapidly changing elemental space and its social representations, characterized by hybridities that are not very stable or well-defined: coastal Louisiana. This region is highly dynamic: the Mississippi River in particular, with its extensive sediments, has shifted the coastal fringe of present-day Louisiana into the Gulf of Mexico. This land gain is contrasted by natural processes, but also by processes resultant of human intervention which cause marine encroachment. In addition to subsidence processes caused by the deposition of sedimentary material and glacial isostatic compensation movements, the drainage and prevention of extensive sedimentation from policy measures of river regulation coupled with the rise in sea level produced by anthropogenic climate change, for example, lead directly to land loss, which is indirectly exacerbated by the construction of canals and invasive species. Louisiana's land loss makes the question virulent of how to describe emerging hybrid spaces and highlights the limits of a positivist understanding of boundaries that is also physically geographical. In the neopragmatic tradition, positivist research findings will be framed in social constructivist terms and supplemented by phenomenological approaches to Louisiana's coastal space, thus suggesting the need for and potentials of horizontal geographic integration of different theoretical and methodological approaches as well as researcher perspectives and data bases.

Contents

1 Introduction

The topic of the coast, especially with its special aspect of ‘beach’, receives particular attention in cultural and social studies (cf. Denzer et al. 2010; Döring et al. 2005; Kühne, Berr, and Jenal 2022; Lenček and Bosker 1998; Osbaldiston 2018; Preston-Whyte 2004; Urbain 2003). They represent landscape segments that can be considered ‘abnormal’ or ‘hybrid’ both in terms of their material foundations and in terms of symbolic associations as well as the patterns of activities located there (Fiske 2003; Kühne and Weber 2019). What can be stated for the coast in general can be found particularly intense as regards the Louisiana coast. The coastal region of Louisiana can be described as a physical as well as social space of fivefold extremes: First, is characterized by extreme physical dynamics, through the processes of land deposition and land loss. Second, it is subject to extreme human activity, some of which is neither immediately nor equally intense in all places. Third, these extreme dynamics require responses of local societies that far exceed autochthonous adaptive capacities. Fourth, and most relevant from a scientific perspective, these dynamics pose extreme challenges to scientific disciplines in analyzing them and presenting the results. These four extreme aspects lead, in turn, to the fifth—policy responses very much inclined to the extreme. These five extremes, in their complexity, make the coastal region of Louisiana a very appropriate space to test the appropriateness of theories, methods, data, researcher perspectives, etc., here especially under the metatheoretical framework of neopragmatic theory of horizontal geographies (the outlines of which will be discussed in more detail). To date, land loss in Louisiana has been subjected to extensive study from both natural history plus social and cultural science perspectives (for examples, see: Bernier 2013; Bisschop et al. 2018; Boesch et al. 1994; Britsch and Cunbar 1993; Colten 2018, 2021b; Craig et al. 1979a; Morton et al. 2006; Nittrouer et al. 2012; Olea and Coleman 2014; Templet and Meyer-Arendt 1988). In comparison to these engagements with land loss in Louisiana, our work represents a neopragmatic redescription (in the sense of: Rorty 1991, 1997; see also: Topper 1995). Accordingly, by means of a theoretical and metatheoretical framing that has yet to be

O. Kühne and L. Koegst, *Land Loss in Louisiana*, RaumFragen: Stadt – Region – Landschaft, https://doi.org/10.1007/978-3-658-39889-7_1

substantiated, we seek to correlate what is known and to supplement it with our own empirical surveys. This emerging fabric, in turn, needs to be interwoven back into the theoretical framework. In this way, a redescription of land loss in Louisiana emerges that is cognizant of its contingency.

Our investigation is fundamentally based on the combination of the aforementioned U.S. pragmatism, or in our case neopragmatism, and a research tradition that has its roots in the German-speaking world: This tradition includes Karl Popper with his three worlds theory, Ralf Dahrendorf with his 'life chances' approach, social constructivism in the tradition of Peter Berger, and Thomas Luckmann, while also including the new phenomenology. The different theoretical approaches, placed in the neopragmatic framework, have served to examine different aspects of the material, individual, and social worlds, as well as the relationships between them. In doing so, we draw on the tradition of landscape studies, which has developed numerous theoretical perspectives over the past decades—something we can harness for our inquiry. However, it is not only because of this heuristic practicality that an investigation with a focus on and use of the concept of landscape lends itself. Even if the aforementioned theorists focus on different aspects of the relations between the material, the individual, and the social, they are united by the intersection of normative orientation, which we—following Karl Popper (2011[1947])—understand as a defense of the 'Open Society'. A society in general, and science in particular, which strives to find suitable solutions for (distinct) challenges, i.e., which does not follow discursive exclusionism or the belief in teleologies (more: Kühne, Berr, and Jenal 2022; Kühne, Berr, Jenal, and Schuster 2021). Following this principle, we specify the basic understanding underlying this book as follows: Karl Popper (Popper 1963; 2011[1947]) understands this, like Max Weber (2010 [1904/05]) and later Peter Berger (2017[1963]), as the essential task of the social sciences in general, and of sociology in particular, to investigate unintended consequences of human actions. If this task is translated into social science research concerning spatial aspects, the challenge arises—wherein non-social construction processes of space are investigated—at the object level of integrating the social and the physical world into one research framework at the meta-level of incorporating different research perspectives, such as positivist (prevalent in the natural sciences) and constructivist (prevalent in the social sciences). Our book aims to further develop such a theoretical framework and put it to test utilizing land loss in Louisiana.

Compared to more common areas, landscape is more saturated with patterns of interpretation, categorization, and valuation. As expressed by Max Weber (1972 [1922]), 'landscape' is associated with a subjectively intended meaning. This means that it is attributed a high socially shared relevance, contains preconceptions, biases, and symbolic charges. The individual or social construction of landscape can be understood as a reduction of complexity and complicatedness, on the one hand, and the increase of contingency on the other hand (see under many: Ipsen 2006; Kühne 2019b; Kühne, Edler, and Jenal 2022a; Papadimitriou 2021). 'Complexity' here denotes the diversity of functions of a system, while 'complicatedness' denotes the diversity of structures; 'contingency', in turn, is a designation of that which is neither necessary nor impossible (Müller 2013; Papadimitriou 2020; Ropohl 2012).

The aim of our investigation is to test and further develop the neopragmatic approach of horizontal geographies. This is done contemplative of a rapidly changing physical space and its social representations, which are characterized by hybridities and are not very stable and well-defined. Resultant upon the empirical results obtained, the theoretical foundations used will also be expanded and fine-tuned.

The Louisiana coast has changed significantly since the last glacial period. Mississippi sedimentation processes were joined by glacial isostatic and eustatic movements, the compaction of sediments, the dynamics of salt dome formation and dissolution, all of which were repeatedly modified especially by the impacts of hurricanes. With increasing intensity, however, man also intervened and still intervenes in the dynamic processes surrounding the development of the Louisiana coast, not least through the imposition of levees upon the Mississippi River (and its tributaries), the construction of canals, and the extraction of raw materials—above all oil and natural gas. The result is a hybrid region that is characterized by processes of different 'cultural-natural hybridity' (Kühne 2012a) which cannot be understood in its complexity by scientific methods alone, and which is changing as a result of its rapid transformation (Colten 2021b). It is difficult to categorize it unambiguously because of its rapid changes. This ambiguity, in turn, offers reason to deal with the questions of the social production of landscape as well as its medium, especially cartographic, of construction. This examination is carried out under the metatheoretical framework of neopragmatic theory according to Richard Rorty (Rorty 1982, 1991, 1998), in its transfer to both 'horizontal geographies' (Kühne 2018e; Kühne and Jenal 2020b; in the most current development: Kühne 2022c) and, contrastingly, the theory of three worlds from Karl Popper (Popper 1979; Popper and Eccles 1977) from which is derived the theory of the three landscapes (Koegst 2022; Kühne 2020a; Kühne and Jenal 2020c; Kühne, Koegst et al. 2021). This metatheoretical framework simultaneously allows a differentiated investigation of landscape according to social contents, individual conceptions, and materialities (theory of three landscapes), which is then appropriately processed from the research question with the help of different theories, empirical methods, researcher perspectives, data, modes of landscape construction, as well as forms of representation (neopragmatism). Starting from this metatheoretical background, we will address an overview of the current state of research regarding different processes of formation and loss of land along the Louisiana coastline. Following this, we will address the problems of representing coastal regions characterized by hybridizations and ambiguities. After conceding the failure of those attempts to precisely define the hybrid areas of coastal Louisiana, we will devote ourselves to their convergence with a scientific approach to a method of spatial study consistently opposed to the positivist understanding of the world: the phenomenological experience of land loss. In conclusion, we will demonstrate that a multi-perspective neopragmatic approach to complex questions serves to provide a more comprehensive understanding of the world—not least its contingency.

2 A Neopragmatic Theoretical Framing and Its Operationalization

This book aims to test, further develop, and operationalize the neopragmatic approach of horizontal geographies involving the issue of land loss in Louisiana plus to further sharpen certain theoretical approaches against the background of empirical results. The development of the neopragmatic approach can be understood as a product of dealing with complex objects (such as landscapes or regions). In part, it is based on the realization that single perspectives (but also methods, data, forms of representation, etc.) do not do justice to the complexity of the object but require multi-perspective approaches. Concurrently, it is also based on the realization that there are sometimes considerable competing interpretations, even compatibility, between these theories. This can hardly be managed from the perspective of these theories. In this respect, neopragmatism represents a metatheoretical approach to combine different theories—from the research topic and the questions arising from it—in a well-founded way and against the background of their expected suitability for gaining and making available newer descriptions of the world. Thus, in this chapter we turn to the fourth extreme mentioned in the introduction, creating an analytical and theoretical framework with which to confront the complexities that develop not least from the other four extremes.

To this end, we will first address the exposition of the complexity of the topic of space and landscape by examining the theory of three landscapes before introducing the neopragmatic metatheoretical framework. The rapidly and fundamentally changing nature of a material space such as the Louisiana coast, not the least of which is human activity, exhibits differentiated and likewise rapidly changing social effects that are challenging to evaluate. For us, the basis of this evaluation is the maximization of opportunities in life ('life chances'), which we present before the methodological operationalization of the neopragmatic approach. These, in turn, are the basis for the design of our study, which we present at the end of this chapter.

O. Kühne and L. Koegst, *Land Loss in Louisiana*, RaumFragen: Stadt – Region – Landschaft, https://doi.org/10.1007/978-3-658-39889-7_2

2.1 Landscape 1, 2, and 3—Some Basic Features

Karl Popper's theory of three worlds (Popper 1979, 1984; Popper and Eccles 1977; for general introduction: Alt 1995; Franco 2019; Niemann 2019) has already been considered by different authors regarding the discussion of spatial understandings (Hard 2002; Schafranek et al. 2006; Weichhart 1999; Werlen 1986, 1997; Zierhofer 1999, 2002). However, in the face of the poststructuralist or critical human geography mainstream, it has not been able to establish itself permanently (Korf 2021). More recently, however, this theory has become increasingly established in social science landscape research (Gryl 2022; Koegst 2022; Kühne 2018a, 2020a, 2021d; Kühne, Berr, Jenal, and Schuster 2021; Kühne and Edler 2022; Kühne and Jenal 2020c; Kühne, Jenal, and Edler 2022).

In his 'Theory of the Three Worlds', Karl Popper distinguishes a material World 1, from a World 2 internal to individual consciousness, and both from a World 3 of cultural content. Following this, 'the theory of three landscapes' can be derived equivalently: Landscape 3 comprises those contents of World 3 that contain social conventions and conventions solidified into culture for the synthesis of phenomena into landscape (these are landscape-related patterns of interpretation, valuation, and categorization). Landscape 2 comprises the respective individual conceptions of landscape, whereby these are based on individual experiences of landscape contents that have been interpreted from World 1. Contrastingly, these experiences are synthesized within the consciousness by the acquired conventions for landscape synthesis. Thus, they are significantly shaped by Landscape 3. Landscape 1, in turn, comprises those objects of World 1 that (socially, i.e., based on Landscape 3, or individually, based on Landscape 2), are thereby synthesized into 'landscape'. Following this view, Landscape 1 cannot be perceived independently from Landscape 2 or 3. Landscape 1 arises exclusively when material objects of World 1 are subjected to a synthesis to landscape (as Landscape 1). Like the three worlds, Landscapes 3 and 2 are connected exclusively by Landscape 2. Landscape 2 mostly confirms the acquired patterns of interpretation, evaluation, and categorization of Landscape 3 and is able to change through innovation involving Landscape 3. Landscape 2 subjects Landscape 1 to observation, Landscape 1 structures Landscape 2, and Landscape 3, as indicated, provides Landscape 2 with the interpretive and evaluative patterns to landscape (which in turn can be accepted or rejected). In the theory of the three worlds (as well as of the three landscapes) man takes a hybrid position, because he has a share in all three worlds (also landscapes), since he has a body (World 1) and a consciousness (World 2) and the latter in turn has elements of World 3 (cf. Fig. 2.1). The human being has therefore within his 'inner self' as in the sense of a feeling (German: Leib), experiencing, and perceiving body a share in World 1 and has elements of World 3 inscribed within this 'inner self'/Leib (something in the form of the habitus; further see: Bourdieu 1984; Brüntrup 1996; Bunge 1984; Deffner and Haferburg 2014; Dörfler and Rothfuß 2018; Knoblauch 2003; van Essen 2013; Waldenfels 2000; Zoglauer 1998).

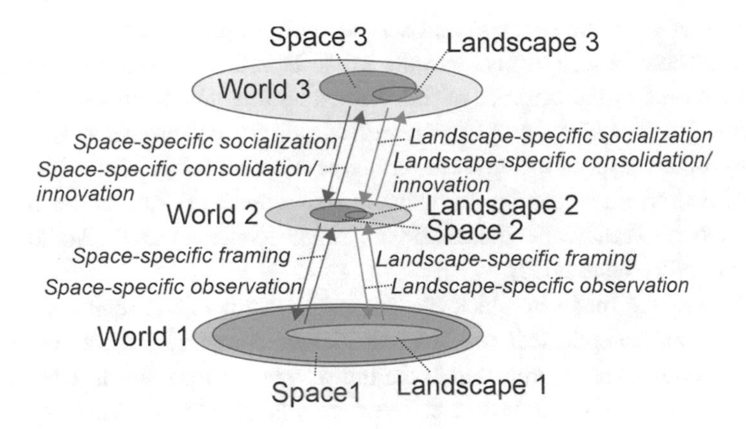

Fig. 2.1 The Worlds 1, 2, and 3 in their relations to their spaces and Landscapes 1, 2 and 3. It becomes clear that space and landscape are not congruent, but especially landscape on levels 2 and 3 can also show non-spatial metaphorical contents. In our investigation, we will focus on those landscape contents that are also subsets of the spatiality. (Own representation)

The construction of Landscape 2 does not follow a uniform pattern. In addition to cultural differentiation (see for instance: Bruns et al. 2015; Drexler 2013; Makhzoumi 2002), it takes place in three different modes (Kühne 2021d; Kühne, Edler, and Jenal 2021a, 2022a): Mode A denotes the construction of landscape as a 'native normal landscape', which is formed in childhood through the individual experience of World 1 as landscape, mediated by mostly close relatives. Landscape in Mode B harkens back to the common perception of socially shared landscape stereotypes. These are formed on the basis of the contents of mass media, the relevant classroom lessons, of books, illustrated books, brochures, etc. They include, in particular, both aesthetic and ecological patterns of interpretation, valuation, and categorization—those that can be presented in a relevant social context without loss of social recognition. The construction of landscape in the Mode C, in turn, is the result of landscape-related occupational aspects. The Mode C is characterized by 'expert special knowledge'. These, in turn, are highly differentiated in terms of expertise (for example, the Landscapes 1c of a landscape ecologist, a curator, conservationist, a forester, and an agricultural economist differ significantly; Hunziker et al. 2008a, 2008b; Hunziker 2010; Kühne 2006a, 2018d; Stotten 2013). Connected to the different modes are also different normative demands on the World 1 (or Landscape 1) interpreted as landscape. In Mode A, the norm of stability is addressed to Landscape 1, which means 'native normal landscape' and does not have to be stereotypically beautiful, or even particularly ecologically valuable, rather it has to be familiar. In contrast, the common perceptions which Mode B searches for in Landscape 1 are the stereotypical aesthetic and ecological notions of landscape that shape this Mode B. The norm of Mode C, in turn, is shaped by the standards of the various disciplines, which not only compete with each other, but also with Modes B and A. Moreover, they are subject to the dynamics of scientific progress and subsequently to the undermining of established perspectives. These in turn diffuse into the Mode B and via inscription in Landscape 1 also into Mode A (for the empirics of this theoretical framing, cf: Böse et al. 2019; Jenal 2019; Kühne 2018d; Leibenath 2014; Stemmer et al. 2019; Stemmer et al. 2020; Weber et al. 2018). In this respect, the effort to further develop landscape observation in Mode C can also be interpreted as striving for gains in distinction in Bourdieu's sense (Bourdieu 1984; in relation to landscape: Kühne 2006b, 2008; Schneider 1989; Wojtkiewicz 2015; Wojtkiewicz and Heiland 2012).

Regardless of the mode in which landscape is constructed, it represents a reduction in complexity and complicatedness. The reduction in complexity occurs because not all elements of World 1 are incorporated in the transformation from World 1 to Landscape 1 (for example, not every single leaf or blade of grass is added to the landscape synthesis). The reduction to landscape occurs in a twofold manner: First, 'complexity' is reduced at the level of Landscape 1 by synthesizing (for example, blades of grass, leaves, flowers, stems, etc. into meadow), relations between certain individual elements (for example, between a blade of grass and a flower) are not assigned to any single consideration. Secondly, Landscape 1 is constructed from certain functional modes of Landscape 2 and 3, the synthesis takes place under the pattern of interpretation, and connected to this, also of evaluation—for instance, as natural landscape, recreational landscape, industrial landscape, etc. This selective complexity reducing construction, in turn, can contribute to an insight into contingency, provided that alternative interpretations and evaluations of landscape are recognized as legitimate (Kühne, Berr, and Jenal 2022; Kühne, Edler, and Jenal 2022a).

2.2 Neopragmatic Landscape Theoretical Approach

From the previous chapter, the diversity of different perspectives in Mode C become clear, as well in terms of their distinctions from the Modes A and B. Following a modernist logic of dealing with this diversity of perspectives, there arises a competition for interpretive sovereignty, as a hegemony or 'colonization of lifeworld landscapes' (i.e., the colonization of Modes A and B by Mode C; for more detail, see: Aschenbrand 2016; Ellmers 2019; Haberl 1998; Kühne 2008; Poerting and Marquardt 2019; Wojtkiewicz 2015). Neopragmatism takes a different path here.

Karl Popper's 'spotlighting hypothesis' ('Scheinwerferhypothese' (2011[1947]) states that theories are only capable of illuminating partial aspects of complex research objects. Other partial aspects, however, remain unexplored (Fig. 2.2). Following neopragmatism, in particular Richard Rorty (Rorty 1982; 1997; Rorty 1998), a framework is developed for integrating different theoretical approaches (understood as 'spotlights') into a research program (for detailed introductions to neopragmatism, see: Gustafsson and Brandom 2001; Menand 2001; Mounce 2002; Müller 2021; Reese-Schäfer 2016).

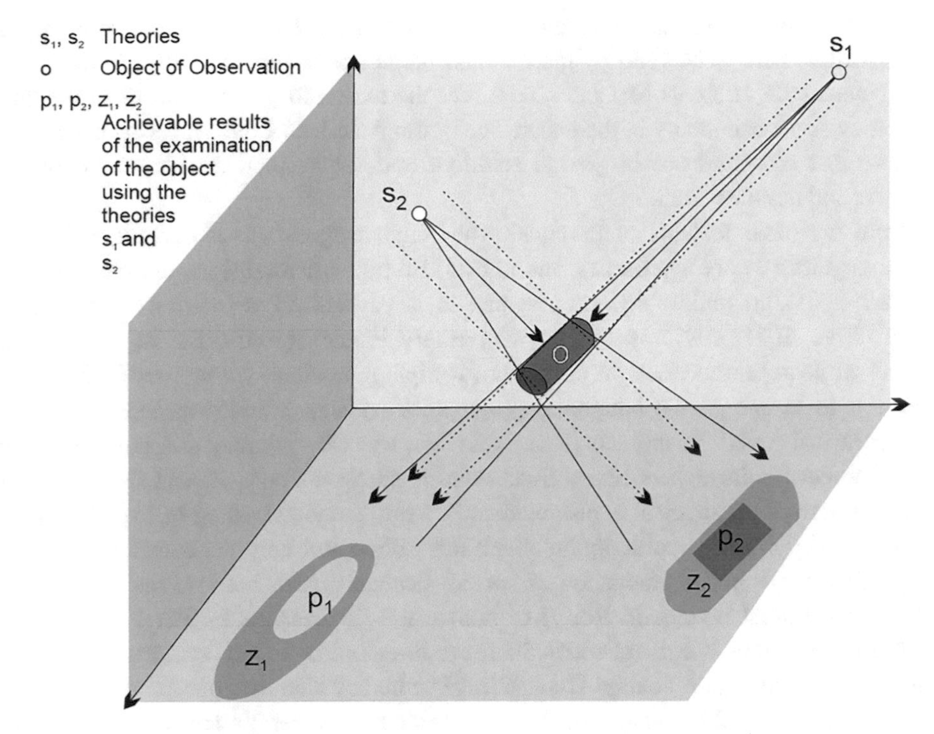

Fig. 2.2 Visualization of the need for theory pluralization. (Own representation according to: Hügin 1996)

(Neo)pragmatic considerations certainly resonated in human geography (esp: Barnes 2008; Cutchin 2008; Jones 2008; Peet 1998; Sayer 2010; Steiner 2009, 2014), but to date there has been a lack of systematic elaboration of a synthesized 'horizontal geographic' research program (i.e., in relation to regional geographies, the study of landscapes, individual settlements, etc.). Only one such study has taken place since late in the second decade of the twenty-first century (for instance at: Chilla et al. 2015; Eckardt 2014; Kühne 2018e; Kühne and Jenal 2020a; Kühne, Parush et al. 2022; Stemmer et al. 2022; Zepp 2020). Thereby 'horizontal geographies' (Kühne 2018e) are, because of their complex subject matter, especially suited for a multitheoretical synthesis.

The foundation of neopragmatism is American pragmatism as in a 'consciousness to action' (Gethmann 1987). Pragmatism advocates the primacy of practical criteria over abstract theory. The criterion for 'truth' is the success of ideas, "interests, values, and goals" (Barnes 2008, p. 1544). By whatever "truth happens to an idea. It becomes true, it is made true by events" (James 1907, p. 67), which means the elision of normative/moral meaning ('good' vs. 'evil'; see Gloy 2004). The pragmatic concept of truth is specified by John Dewey with his concept of 'warranted assertibility' (Dewey 2016; see Neubert 2004): That which is called 'true' is located at the intersection of that which has already been worked out by means of scientific methods and results interconnecting with that which will be held to be valid by means of methods and results developed in the future (vgl. Neubert 2004). Louis Menand (2001) sees the reason for the emergence of pragmatism as an American theory in the experience of the American Civil War. This experience led to a loss of confidence in general solutions and undermined the ideas of universal progress and absolute truth.

From the basic features of pragmatic thinking emerges that of neopragmatism, in which linguistic theory approaches, and in part also poststructuralist approaches, are used (Baltzer 2001). In philosophy, neopragmatism is particularly associated with Richard Rorty (1982, 1991, 1997) as well as with Hilary Putnam (1995). Barnes outlines the sources of neopragmatic thought of Richard Rorty, upon whom we primarily rely in our approach to 'neopragmatic horizontal geographies' (2008, p. 1543) as follows: "Apart from the usual American suspects (Rorty's favorite was Dewey), they also included in his view European philosophers like Wittgenstein, Heidegger, Foucault, and Derrida". The linguistic turn of pragmatism is also evident in Rorty's understanding of 'truth': Instead of 'truth', Rorty posits 'redescription'. With this notion, not only the constitutive importance of language in the production of 'world' becomes clear, but also the contingency (one of the central notions in Rorty) of human life (Baltzer 2001). For Rorty, neither knowledge of nature, nor moral convictions, nor insights into social facts are to be understood as a reflection of a 'reality' (Baltzer 2001) whereby also the concept of 'discovery' becomes obsolete and is replaced by that of 'invention' (Baltzer 2001; Rorty 1991, 1997). Rorty replaces fundamental differences such as lie/truth or appearance/reality (1994) by an understanding of useful/useless as helpful ways of speaking. At this point he follows Charles Darwin on the one hand, John Dewey on the other. The first in that he describes not so much the knowledge of 'reality' being the task of science, but the adaptation to that reality. He follows the second one in its instrumentalism, ideas are understood here as instruments or tools with which humans are empowered to access the 'world' in order to solve problems (Dewey 1996; Rorty 2001b; this figure of thought can also be found in Karl Popper 1996; cf. also Hagendorff 2011).

As mentioned, in Rorty's philosophy the concept of contingency as that which is neither necessary nor impossible is of central importance. The concept of contingency stands thereby "at one end of a set of oppositions whose other term is *variously necessary, essential, intrinsic,* or *unconditional*" (Topper 1995, p. 958; emphasis in original). The 'insight into contingency' extends beyond philosophical reflection: "For it is not enough to recognize that there is a multiplicity of other vocabularies besides one's own; rather, one must relate this observation to antifundamentalist conceptions of language, man, and the world in order to be able to make the diagnosis of contingency" (Deines 2008, p. 158). With the concept of 'vocabularies', following Wittgenstein's notion of the language game (Wittgenstein 1995 [1953]), Rorty captures (1991) common platforms of social practices of justification. Justification here always refers to the context of the vocabulary of a specific language community (Müller 2021). Since the language communities (including scientific ones) are different, a need for justification of one's own approaches to the world arises (Rorty 1981). Rorty does not understand 'justify' as 'prove' or 'substantiate', rather he understands 'justify' in the sense of "unfold reasons that speak for this conviction" (Rorty 2001a, p. 259).

Such manifestations of the awareness of the contingency of the self, society finds voiced in the embodiment of the figure of the 'liberal ironist' who epitomizes Rorty's democratic antifundamentalism: This person faces the fact "that her central beliefs and needs are contingent" (Rorty 1989, p. 14). In this context, the function of 'good' redescriptions, i.e., 'suitable', becomes clear: They "help to reveal the sheer contingency of our concepts and perspectives. Once a person recognizes the contingency of her inherited final vocabulary and other final vocabularies, she is more inclined to experiment with different descriptions or conceptions of herself and others and alternative ways of life" (Frazier 2006, p. 462 f.). These redescriptions become necessary when the forms of argumentation of a 'time-honored' vocabulary have become questionable to the extent that the claim to suitability can no longer be honored (Rorty 1997). 'Redescription' should not be misunderstood as an attempt to present 'the right description'. Rather, it is an attempt to deal with unsolvable problems, irregular conflicts, or irresolvable anomalies. Redescription thereby create a 'fabric' that weaves in and around 'time-honored' vocabularies creating a new vocabulary (Topper 1995). This new fabric thereby recycles the remnants of those vocabularies it replaces (Rorty 1997; Topper 1995). The vocabularies newly created by redescription are—if they follow the logic of the 'liberal ironist'—not characterized by the striving for hegemony, but rather an expression of the awareness that also the redescription vocabulary in its suitability to interpret 'reality' is both an expression of contingency, and ensuingly can always serve as a basis to be rewoven in its own part into an even newer and more suitable vocabulary (Rorty 1997; Topper 1995).

In sociopolitical terms, neopragmatic thinking corresponds to an open society, which Rorty understands quite normatively with its open-ended, democratic processes of negotiation (see more precisely: Hildebrand 2003, 2005; Müller 2021; Rorty 1989; Warms and Schroeder 1999). Although the awareness of contingency has also shattered the illusion of the ultimate justification of democracy, elementally the democratic polity is also only a contingent and (probably) provisional result of cultural history (Rorty 1989). In his understanding of society, Rorty follows the negativist understanding of liberal democracy of Judith Shklar (2020)[1]. In this reading, liberalism is oriented toward the prohibition of interfering with the freedom of others; what it does not have, conversely, are positive doctrines about how people should lead their lives or even what personal choices they should make (Shklar 2020).

A concept that provides an operationalization of these considerations is that of 'life chances' by Ralf Dahrendorf, with which he traces the contextualization of the human

[1] This view is also represented by Karl Popper and Ralf Dahrendorf with their understanding of 'democracy as change of power without bloodshed'; cf: Dahrendorf (1983); Popper (2011[1947]).

being between individual preferences and social bondage and normatively grasps it in the sense of an open society. We will deal with this concept in the following.

2.3 Life Chances as a Framework for Normative Statements

Ralf Dahrendorf conceives 'life chances' as "first of all chances of choice, options. They require two things, entitlements to participation and an offer of activities and goods to choose from" (Dahrendorf 2007, p. 44). The chances of choice must be connected with a meaning for the person acting. Acting is again understood in the sense of Max Weber (1972 [1922]) as human behavior engaged in external or internal doing, yet also as omitting or tolerating of action, if and insofar as the person or persons acting connect a subjective meaning with it. The basis for Ralf Dahrendorf's understanding of life chances is that of Max Weber, who understands 'life chances' as chances that a person can reasonably expect in his or her life (Weber 1976 [1922]; see also Berger 2017[1963]), with which he initially references to the unequal distribution of life chances in society. Also in his understanding of chance, Dahrendorf concludes (1979, p. 98; but also Dahrendorf 1968, 1992) in conjunction with Max Weber (1972 [1922]): He grasps it, on the one hand, as a "structurally based [...] probability of behavior," and, on the other hand, "as something that the individual can have, something as a chance of satisfying interests" (Dahrendorf 1979, p. 98). Life chances are—both for Dahrendorf and as previously with Weber—dependent on social contexts: They are "possibilities for individual growth, for the realization of abilities, desires, and hopes, and these possibilities are provided by social conditions" (Dahrendorf 1979, p. 50). The influence of society is twofold, society potentially makes life chances possible, for example through education; conversely, it also restricts life chances. This becomes particularly clear in the example of class differences, when "the opportunities of some (the 'ruled', the 'dependent') are decisively predetermined by the choices of others ('the rulers')" (Niedenzu 2001, p. 178).

Ralf Dahrendorf's concept of life chances is based—here he goes beyond Max Weber's understanding—on two concepts, because "life chances are opportunities for individual action that result from the interrelation of options and ligatures" (Dahrendorf 1979, p. 55). In this context, options are "choices given in social structures, alternatives of action" (Dahrendorf 1979, p. 50). These require "choices and are thus open to the future" (Dahrendorf 1979, p. 108). By ligatures, comparatively, Dahrendorf understands "structurally marked fields of human action. The individual, by virtue of his social positions and roles, is placed in bonds or ligatures" (Dahrendorf 1979, p. 51). Ligatures are often emotionally and morally charged and provided with social obligation. However, they also produce affiliation, which allows them to be understood as "foundations of action" (Dahrendorf 1979, p. 51). Under ligatures, Dahrendorf includes things like "the ancestors, the homeland, the community, the church" (Dahrendorf 1979, p. 51). This highlights a fundamental difference between ligatures and options: "ligatures are prerequisite, options are desired" (Dahrendorf 1979, p. 108). In this context, the emergence of life chances is constitutively tied to the existence of both options and ligatures: "ligatures without options mean oppression, while options without ties are meaningless" (Dahrendorf 1979, p. 51 f.). While for Dahrendorf options always have a positive influence on life chances, ligatures remain ambivalent for him. While ligatures turn mere opportunities into "opportunities with meaning and significance, that is, life chances" (Dahrendorf 1979, p. 51), they also adversely have "an absolute quality: they only reluctantly permit shades of gray. People either belong or they don't, and if they don't, they have no claim to rights" (Dahrendorf 1979, p. 51).

Following Karl Popper's dictum that history 'in itself' has no meaning, meaning must rather be given to history, Dahrendorf (1979, p. 26) expresses this in terms of "creating more life chances for more people". This in turn presupposes the individual ability to recognize the differences between options and ligatures in terms of the associated consequences (cf: Kühne, Berr, Schuster, and Jenal 2021). This is all the more necessary because options and ligatures change, additionally, options can also be transformed into ligatures. The increase in life chances, for example, is based on the gain of options, which often resulted from "the breaking of ligatures" (Dahrendorf 1979, p. 52). But also in another direction, powerful people vis-à-vis 'less powerful' people (Paris 2005) are striving to secure their options in the form of privileges and to convert these privileges into ligatures for the rest of society. This has the side effect of limiting the generation of options for the rest of society (Dahrendorf 1992, 2007; Strasser and Nollmann 2010).

Following the previously expressed understanding of Karl Popper's three worlds (cf. Sect. 2.1), World 2 is at the center of Dahrendorf's reflections upon life chances: Options are something that World 2 has to fight for in confrontation with World 3. Ligatures are elements that are first brought to World 2 by World 3. Thus, for Dahrendorf, World 3 represents in large part the "vexatious fact of society" (Dahrendorf 2006, p. 21). We consider this understanding to be both insufficiently differentiated (we address this below) and in need of expansion (to be addressed in Sect. 5.2).

Dahrendorf clearly accentuates Max Weber's concept of life chances with the differentiation into options and ligatures and the expansion of the definitional element, "furthermore, it becomes clear in a theoretical perspective that […] Dahrendorf assumes a socially structured elective action among social actors" (Mackert 2010, p. 413). Nevertheless, the concept of ligatures—especially in comparison to options—remains insufficiently encompassing and difficult to operationalize, which is largely an aspect of their conceptual double function of giving meaning to options yet just as possibly preventing options: Whereby the limiting function of ligatures receives a far more detailed inscription. When Dahrendorf links modernization as an expansion of life chances to the—already quoted above—"breaking of ligatures" (Dahrendorf 1979, p. 52) for example, when social as well as spatial mobility means "that the family and the village are no longer communities of fate, but increasingly become communities of choice" (Dahrendorf 1979, p. 52), this illustrates the asymmetry in Dahrendorf's concept of ligatures in favor of option reduction.

Based on considerations of both the insufficient capacity for complexity and the reduced ability to be operationalized in Ralf Dahrendorf's concept of ligatures, Kühne et al. (2022) make a threefold differentiation of ligatures: First, involving moral and ethical ligatures; second, involving internally directed and externally directed ligatures; third, involving implicit and explicit ligatures. Moral ligatures present unquestioned norms of action to the World 1 (such as religious norms), while ethical ligatures are meta-level instruments to question and weigh moral ligatures. Ethical ligatures thus provide a basis for maximizing individual life chances in an open society, while moral ligatures constrain those individual life chances through 'thou shalt' demands deemed unquestionable. Ethical ligatures include such principles as procedural justice, tolerance, responsibility, enabling contextualized thinking, and so on. They can also be understood as 'higher order consensuses' (in Nida-Rümelin's sense) that "align with procedures, the manner or method of collective decision-making" (Nida-Rümelin 2020, p. 114). Internally directed ligatures are directed toward actions of World 2, externally directed ligatures are directed toward the actions of any other World 2; here, considerations by David Riesman (1950) flow on internally and externally directed character. Externally directed moral ligatures are especially able to cause individual and social conflicts—outcomes of often incommensurable morals—which are difficult or impossible to regulate (Berr et al. 2022; Kühne 2019a; Kühne, Schönwald, and Jenal 2022). Ethical internally directed ligatures, on the other hand, enable individual consideration and adherence to values and norms, without any claim to generalization. Following internally directed moral ligatures means the largely unquestioned following of the moral norms applied to World 2, while externally directed ethical ligatures expect the generalization of a meta-level reflection of moral ligatures (which in turn also holds considerable potential for conflict, for example, if the willingness to reflect ethically on morality is not given). Explicit ligatures are formulated openly (such as laws, ordinances, instructions for scientific work, etc.), while implicit ligatures are not, but are tacitly assumed and thus offer a high potential for distinction; In many cases, they have become part of the habitus. However, implicit ligatures can be made explicit (a task of sociology as well as other social and cultural sciences, but this also takes place beyond academia, as in books and courses on etiquette). This also means that explicit ligatures are more open to becoming the object of reflection based on ethical reflections.

With this differentiation, ligatures can be assigned the following against the background of the norm of increasing life chances: Ethical ligatures promote individual life chances, moral ligatures restrict them, internally directed ligatures give the World 2 standards for the direction of action, externally directed ligatures harbor conflicts (whereby in each case, with regard to life chances, moral ligatures, regardless of directionality, contribute to restriction, ethical ligatures to enlargement). The more explicit ligatures are, the more likely they can be the object of reflection on the basis of ethical ligatures, and thus contribute to the enhancement of life chances; the more implicit they are, the more likely they contribute to the reduction of life chances.

2.4 Operationalizing Previous Thinking into a 'Horizontal Geography' Research Program on Louisiana Coastal Land Loss

After this brief introduction to central aspects of neopragmatic thinking as well as the normative framework of life chances, considerations of translation into a 'horizontal geographic' research program, in the sense of a redescription, are carried out in the following. As mentioned, subjects of horizontal geographic research are characterized by a high degree of complexity, in the sense of different possibilities of interaction between elements (for more details see: Papadimitriou 2021, 2022). This complexity of the subject makes a single theory approach seem unpromising, particularly when 'unifying theories' or 'lack of contradictions' is not at the center of interest but instead an understanding of the complexity and contingency of the subject are. Redescription refers in this sense to an interweaving of the familiar vocabularies of those differing theories that do not fundamentally contradict the central argument of neopragmatism concerning the contingency of self, society, and the world in general, such as essentialism. In this respect, the approach of 'neopragmatic horizontal geographies' forms a metatheoretical fabric by a recycling of theories which among themselves competed for an interpretative sovereignty approaching incommensurability and insofar can be brought to an 'integrative interweaving' solely via a neopragmatic metatheoretical redescription. This understanding of a redescriptional integration is not limited to a multitheoretical access alone, but through triangulations on an additional five levels (here it becomes clear that the 'neopragmatic approach' presented here goes far beyond a mixed-methods approach).

The redescription of 'neopragmatic horizontal geographies' herein comprises a total of six levels of triangulation. These take the following form (more detail on the different triangulations: Denzin 2007; Flick 2007, 2011; Jick 1979; Kuckartz 2014; Kühne 2021a, 2022c; Kühne and Jenal 2021a, 2021b; Morse 1991; Schründer-Lenzen 2013):

1. *Theory triangulation* is the core aspect of the neopragmatic approach. Here, different theoretical perspectives on a complex phenomenon are used. The choice of theories requires justification against the background of the research focus as well as the research question. The choice of theories requires justification against the background of the expected (contingent) findings.
2. Theory triangulation is followed by *method triangulation.* This, too, needs to be justified in the context of the research topic and the research question against the background of the expected results.
3. Method triangulation is often followed by *data triangulation.* The combination of data from different sources as well as scope and type (qualitative or quantitative) helps to avoid one-sided interpretations. Appropriateness of the type and extent of data in relation to the research question and expected outcome is central here. Also, the

informative value of the data used must be reflected, regardless of whether it is data collected by third parties or data collected in-house.

4. *Triangulation of modes is* another core aspect of the neopragmatic approach. As shown above, the Mode C represents a specific view of the world, connected with specific patterns of interpretation and especially of evaluation. In the sense of a transdisciplinary integration, it is also necessary to integrate the views of Mode A and B into the investigation.
5. In the *triangulation of the forms of representation,* it is important to develop appropriate forms of representation—linguistic, graphic, and cartographic—in relation to the question and the results, the quality of the foundational data, and especially the comprehensibility for potential users: preferably in a form that clarifies the continuity of world.
6. *Researcher triangulation* is also a consequence of triangulations i-v. The neopragmatic approach to researching complex horizontal geographic issues developed here requires a high degree of competence (knowledge, comparisons of implementation) in dealing with theories, methods, data, modes of spatial generation and representations. These demands are particularly high when recourse is made to different disciplinary logics, such as the natural, social, or cultural sciences. Depending on the research topic and research questions, theories and methods, data generation and representation, as well as demands on the integration of Mode A and B plus on the avowed users, a combination of researchers possessing different disciplinary provenances is required. In addition to different professional backgrounds, aspects such as gender, spatial origin (Mode A), cultural and social backgrounds, career status, ideological orientation, etc., must also be considered (depending on the research question and topic).

Linked to the need for justification of the different triangulations is the criterion of the appropriateness of the approach. This means that the investigation should be designed in such a way that it does not show any over-complexity or under-complexity. In the case of method triangulation, for example, care must be taken to ensure that the methods used follow from the theoretical considerations and serve to answer the research question, also that the data triangulation—following the large availability of data—does not contain any evaluations that are unnecessary thematically or in relation to answering the research question. Conversely, it is important to avoid having a neopragmatic approach serve as an excuse to selectively use sparse data.

Grounded in this neopragmatic investigation, we correlate the developments to the understanding of life chances presented above, against the background of the norm of their expansion.

2.5 Conceptual Considerations for our Study

Our study presented here addresses the issue of land loss in Louisiana, focusing on questions surrounding their representations. It is, thusly, a thematically focused horizontal geographic investigation. The thematic focus, in turn, features natural science aspects (especially climatological and geological/geomorphological) as well as social and cultural science aspects wherein media representations or individual experience of the Landscape 1 in question are addressed. With respect to the six levels of triangulation discussed in the previous section, we proceed as follows:

1. Theory triangulation: The neopragmatic conception of the study combines engagement with the natural sciences as well as social and cultural science questions. In addition to neopragmatism, the metatheoretical framework is provided by the theory of three landscapes to analytically organize the levels of landscape while concurrently providing an approach for assigning which of the theoretical perspectives are meaningfully drawn upon. The natural science studies on which our work is based investigate the topic of land loss from a positivist perspective. Since this—as will be shown—fails in its claim to produce explicitness, we frame it in social constructivist terms. From this perspective, we also address the media constructions of land loss, supplemented by an account of our phenomenological experience of land loss, in order to provide a counterpoint to the positivist effort to produce lucidity (for an overview of the different theories, see: Bourassa 1991; Howard et al. 2019; Kühne 2019b, 2021b; Kühne and Berr 2021; Winchester et al. 2003; Wylie 2007).
2. Methodological triangulation: Our study is essentially based on the implementation of four methodological approaches. 1) The notably natural science basics were prepared by a meta-study (based on the relevant hits in a Google Scholar search). 2) The analysis of cartographic representations is based on a literature analysis cross-referenced with an analysis of maps and orthophotos, on the basis of which the problem of the distinct classification of land and water was devised. 3) The media analysis follows the discourse-analytical investigation of Baum (2021), which is supplemented by current media representations in the wake of Hurricane Ida in 2021. 4) The phenomenological reference is carried out through phenomenological walks, which were supplemented by ero-epic conversations with residents of the affected areas (for method: Kühne and Jenal 2020a; Tilley 2008; Wylie 2005). Please note: While 1) is tied back to a positivist understanding of science under social constructivist framing, methods 2) and 3) are clearly derived from the social constructivist perspective. This can also be found in 4), even though the phenomenological walk, a method rooted in phenomenology, does dominate here.
3. Data triangulation: The data obtained by means of methods 1) and 2) are mainly quantitative data, as are their graphical and cartographic processing, which will be interpreted in a social constructivist way. The data used in 3) are especially qualitative,

as well as those of 4), which were created by means of note-taking in the research process or as a memory protocol.

4. Triangulation of modes: In parts 1) and 2), of the study, the Mode C dominates. Especially in part 3) of the study, Mode B perspectives are updated also. Mode A is considered in part 4) of the study, especially regarding ero-epic conversations.
5. Triangulation of the forms of representation: In order to present the diverse and complex processes surrounding land loss in Louisiana in a comprehensible way, graphics, block diagrams, and maps supplement the textual representation. Where the contingency of world could not be represented in the illustrations, reference was made to it in the figure captions.
6. Researcher Triangulation: We conducted the research (supported by research assistants) incorporating different ages, career stages, and gender perspectives as well as professional backgrounds in some instances. The different perspectives were repeatedly addressed in the research process and sometimes brought to light unique points of view.

Now that the metatheoretical foundations of our redescription of coastal land loss have been clarified, as well as its operationalization in terms of triangulations at the different levels, we will look below at the current state of research on coastal land loss, from a natural science perspective in particular.

3 The Multiple Causes of Coastal Land Loss in Louisiana—An Overview

In this chapter, we will apply particular focus upon the first two extremes presented in the introduction directly and indirectly involving coastal land loss in Louisiana, the extreme physical dynamics, through the processes of the formation of land and the loss of land, and the extreme influence of human activities on the loss (as well as on the formation), some of which is neither immediate nor equally intense in all places.

To this end, we will initially elaborate on the topic of coastal land loss and undertake a broader conceptual approach that deals with the hybridity of nature and culture, as occurs in Louisiana and elsewhere. Following this, we will address processes that are dominated by nature, then processes that are influenced or determined by humans. The chapter will conclude with an overview of the different processes in terms of their natural or cultural ties, their temporality, their dimension in relation to Space 1, and the possibilities for mitigation and adaptation by humans.

3.1 Introduction to the Topic of Coastal Land Loss: Dimensions and Further Conceptual Approach

Louisiana's land loss (along with inequality of life chances related to ethnicity and the state's focus on the petrochemical industry with its environmental side effects, among other issues) is considered one of Louisiana's most significant challenges (Colten 2021b; Jenal, Kühne et al. 2021; Kühne 2021c; Kühne, Jenal et al. 2021). Between 1978 and 2000 alone, relative sea level rise in the northern Gulf of Mexico along coastal southern Louisiana represented an annual land loss of approximately 77 km^2 (30 mi^2; Barras et al. 2004). Coastal land loss also has implications for hurricane protection system design, coastal restoration planning, commerce, energy production, and residential developments (Dokka 2011). Land loss in Louisiana is characterized by different causes: Some are found

O. Kühne and L. Koegst, *Land Loss in Louisiana*, RaumFragen: Stadt – Region – Landschaft, https://doi.org/10.1007/978-3-658-39889-7_3

more at the pole of nature in terms of cultural-natural hybridity, others more at the pole of culture. Characterized by a high degree of complexity and reciprocal influence, these processes are, on the whole, interdependent. Those processes which can be assigned to the pole of cultural origination are connected by unintended side effects leading to the land loss of World 1.

The high dynamics of coastal development is not a phenomenon of the present, even if it has acquired its aforementioned extreme nature only in recent decades resulting from human influence (not least climate change). The historical changes between the cartographic representations of 1814 and 2022 are presented in Fig. 3.1. Even though there were, and still are, difficulties defining the classification of land and water areas in both basic maps (detailed in Chap. 4) and the georeferencing is characterized by a certain tolerance (1814 map, therefore a labeling format to express this ambiguity), tendencies in the dynamics of the development of land and sea areas become clear: While in the left third of the cartogram a land gain (notably the Atchafalaya River, south of current day Morgan City) can be recognized and also the mouth of the Mississippi River (according to the underlying map) have led to an increase of land area, massive land losses are indicated in the remaining parts of the Area 1 representation.

In the remainder of this chapter, we will discuss the major process of land loss in Louisiana discussed in the literature as a basis for further elaboration on the social construction and individual experience of coastal land loss (more detailed overviews are provided by, for example: Bernier 2013; Boesch et al. 1994; Craig et al. 1979a; Dokka 2006, 2011; Olea and Coleman 2014; Rogers et al. 2006). The literature evaluated is generally natural sciences in nature and sometimes quite technically scientific in manner and thus follows a (usually implicit) positivist paradigm; cultural and social science treatments of the topic remain in the minority (and tend to treat the significance of constructivist attributions marginally or implicitly). The order of the processes we discuss that lead to land loss is roughly oriented toward first presenting those processes that are

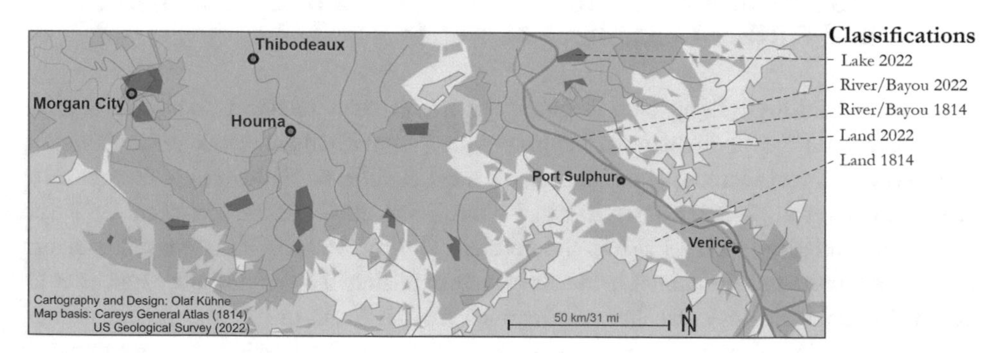

Fig. 3.1 Cartogram showing rough estimates (tolerances in georeferencing) of changes in distribution of surfaces classified as land and as water, comparing 1814 and 2022. (Own cartographic representation after: Carey 1814; U.S. Geological Survey 2022)

more at the pole of nature in relation to cultural-natural hybridity, and then moving on to those that are more at the pole of culture. In addition, there is a tendency in the outline to first deal with large-scale, then later with small-scale processes, moreover, first with processes that cover a large time span and later with those that were or are rather of short temporal duration.

3.2 Processes of Coastal Land Loss Inclined Towards the Pole of the Natural

Isostatic balancing movements are of long-term duration affecting large Space 1 sections and at the pole of nature within a cultural-natural hybridity (Fig. 3.2; cf: Heiland 1992, 2010; Kühne 2012b). Isostasy denotes a hydrostatic equilibrium, which occurs because the relatively rigid lithosphere rests on the relatively fluid and therefore more deformable asthenosphere (Molnar 2015). Equilibrium movements occur during unloading with the consequence of rising (e.g., as a result of ablation) and loading with the consequence of sinking (e.g., by loading of thick ice sheets or delta sediments). Both above processes have significance in the land loss throughout southern Louisiana. Thus, the coastal areas of Louisiana and Texas are affected by postglacial isostatic subsidence. As a result of the high glacial loading during the glacial period, the lithosphere was weighted downward into the asthenosphere across large portions of the northern North American continent. The glacial load to the north caused the more southerly margins of the lithospheric plate to rise isostatically, as in the affected areas of coastal Louisiana and Texas. Accompanying glacial melt was the loss of the downward driving glacial weight load upon the lithosphere. The formerly downward driven portions of the lithospheric plate began to uplift, and the formerly uplifted portions of the plate have since been sinking (forebulge collapse). This process means a subsidence of 80 to 90 m compared to the maximum level (Heinrich et al. 2020). A second isostatic subsidence movement, sediment loading, is also found in southern Louisiana. This is caused by the high sediment load of the Mississippi River. This amounted to 550 million t per year prior to 1950 and 220 million t per year today. Sedimentation of this material carried by the Mississippi River results in mass accumulation. The resulting continuous loading of the lithosphere throughout the Mississippi River delta causes the North American Plate to sink into the asthenosphere—a structure known as the Gulf Coast geosyncline (see Fig. 3.3). Equalizing movements within the asthenosphere, below the Gulf Coast geosyncline, result in uplift approximately along the latitude of Wiggins, Mississippi (30° 51′). Thus, while this area is tectonically uplifted, the area in southern Louisiana is subsiding (Craig et al. 1979b; Dokka 2006, 2011; Heinrich et al. 2020; Mossa 1996; Olea and Coleman 2014; Rogers et al. 2006). As a result of these subsidence processes, the loading with water increases, also as a result of this intrusion the weight adds to the increase of the regional loading of the lithospheric plate,

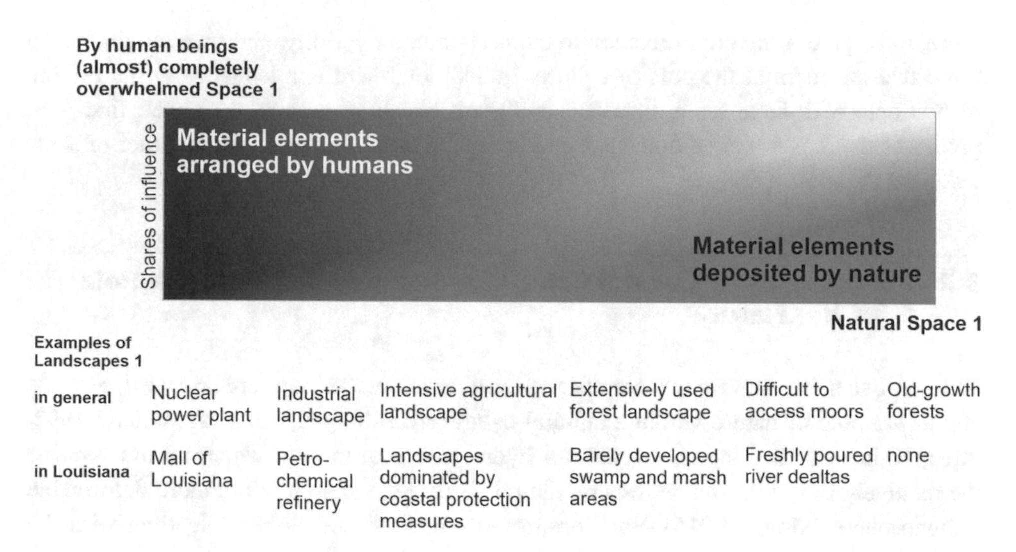

Fig. 3.2 The hybridity of nature and culture at the level of Space 1 or Landscape 1 with general examples and examples from Louisiana. The influence of man on Space 1 in Louisiana is strongly formative, such that it can hardly be assumed anymore in Louisiana that there is not a Landscape 1 unimpacted by human beings, even the freshly filled delta areas are subject to a strong human imprinting by his intensive intervention in the flow regime of the Mississippi River system. (Strongly modified after: Kühne 2012a)

which is correspondingly pushed into the asthenosphere. Some of the impacted asthenosphere, in turn, is displaced toward the north, contributing to the rise of the lithosphere there (Heinrich et al. 2020).

Another process leading to land loss in southern Louisiana is the compaction of sediment. The heavy weight of the thick sediment deposits leads to sediment compaction at great depths, which generates the associated formation of pressure ridges and fold belts (Rogers et al. 2006): Salt and sediment are compressed towards the continental shelf, processes "which are akin to sitting on a peanut butter and jelly sandwich and watching material ooze out and shift" (Rogers et al. 2006, III-37). Sediment uplift also results in the formation of salt domes, the importance of which for land loss is outlined later. The continental slope, as well as the continental shelf, is covered by numerous and large submarine landslides that result in mass displacement away from the coastal area (Craig et al. 1979b; Dokka 2011; Heinrich et al. 2020; McBride et al. 2007; Rogers et al. 2006). These processes are joined by listric thrusting along the coast as a series of faults running roughly parallel to the shoreline. Currently active is the Baton Rouge fault zone, trending east-west along the north shore of Lake Pontchartrain toward Baton Rouge, the state capital. Locally, faults are found around salt domes (Dokka 2011; Rogers et al. 2006; Fig. 3.4).

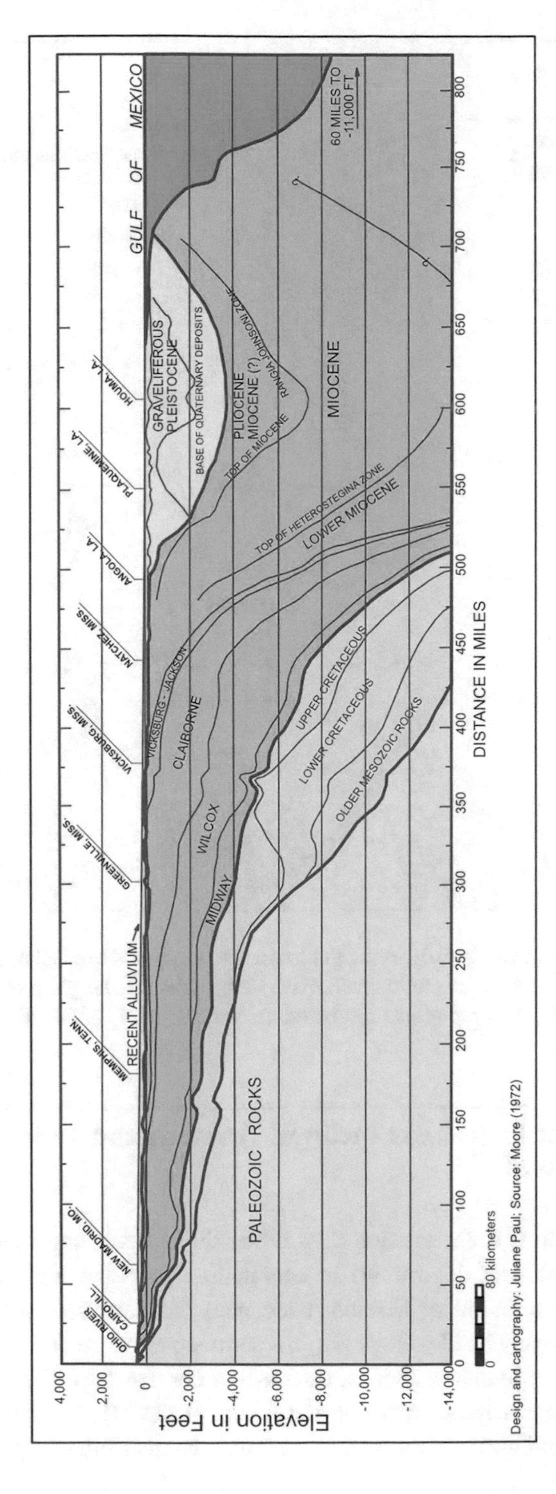

Fig. 3.3 Geologic cross-section of Mississippi River deposits from north to south. (Own representation based on: Moore 1972)

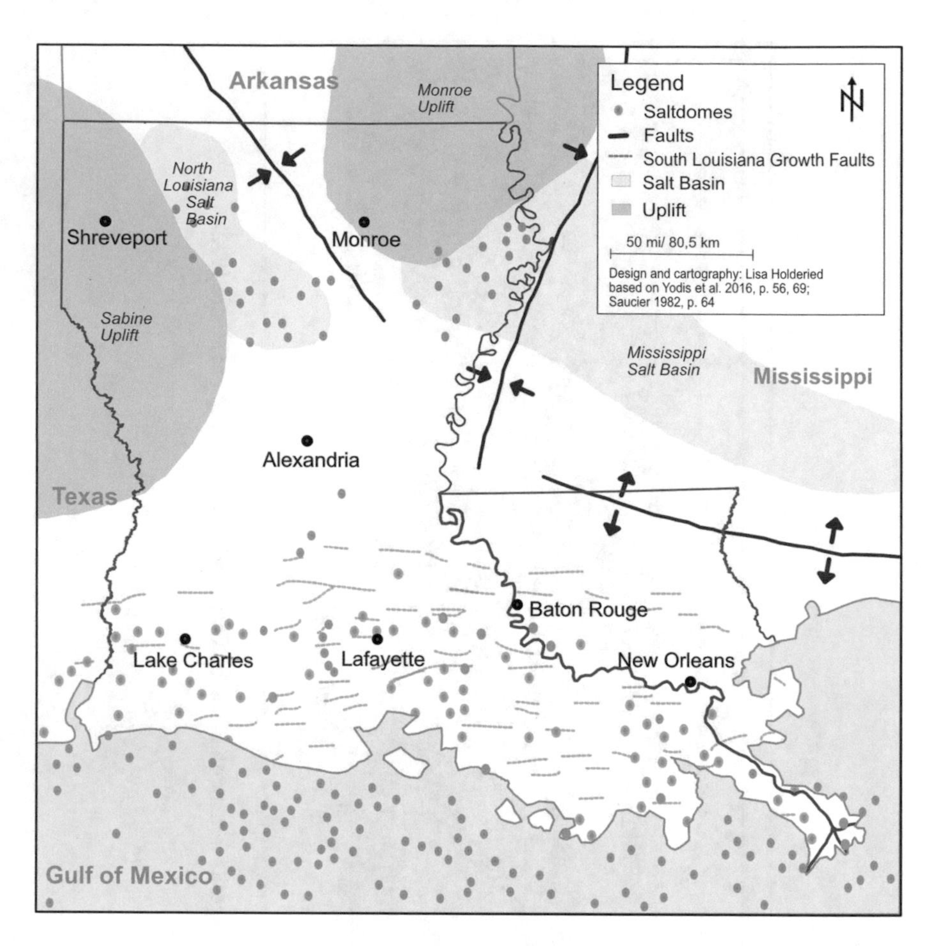

Fig. 3.4 The structural geologic framework of the lower Mississippi River delta, extracted from. Saucier (1994, p. 64)—Growth faults (solid black lines) disturb the coastal plain of the delta as do salt domes. (Shown as dots; own representation based on Saucier 1994, p. 64, supplemented with Yodis et al. 2016, p. 62, 76)

3.3 Processes of Land Loss Inclined Towards the Pole of the Cultural

When the Mississippi River sedimentation does offset the isostatic subsidence processes (also contributed by sediment deposition) in conjunction with the movement of compacted material towards the Gulf of Mexico that causes the Earth's surface to sink, the land surface remains roughly stable. However, this sedimentation process is reduced as a secondary consequence of human activities, such as the construction of dams and levees for flood control. The regularly occurring spring floods of the Mississippi River poses a threat to communities and agricultural land as well as technical infrastructure benefiting the inhabitants of southern Louisiana. While the construction of levees meant a reduction in flood hazards, they also prevented sediment material from being deposited over large areas during flood events. To facilitate navigation on the Mississippi River, the course of the river was straightened, banks were fortified, and jetties were constructed, all of which led to an increase in flow velocity. This flow velocity increased the load capacity of the Mississippi River, resulting in more material being transported to the Mississippi River Delta. Associated with the levee construction is a cutting off of bayous flowing toward the Gulf of Mexico (which can be understood as delta arms) from the Mississippi River. The cutting off not only removed water from the Mississippi River, but also the material transported with it. This significantly reduced the bayou sedimentation capacity in the area, as well as in the estuaries (Beatley 2009). Thus, the regional distribution of sedimentation shifted: in addition to streams with increased estuary input besides the Mississippi River, Tchefuncte River into Lake Pontchartrain as well as Yellow Bayou and the Atchafalaya River, large areas of sedimentary deficiency emerged along the Louisiana coast (Barry 1998; Colten 2021a; Craig et al. 1979b; Khalil 2018, cf. Fig. 3.5).

The desire to make land available for human activities (often agriculture, but also residential and commercial uses as well as infrastructure) has resulted in the drainage of

Fig. 3.5 Erosion control measures on the coast: Breakwaters on the south coast are intended to prevent the erosion of sand, with the almost omnipresent oil production facilities in the background. (Photo: Olaf Kühne 2022)

topsoil. This drainage of topsoil is associated with a reduction in pore pressure, which causes soils to consolidate and settle. Organic soils are particularly affected by this., especially interdistributary sediment parcels, such as the old ridge swamps around New Orleans, are highly compressible. The neighborhoods built here show obvious signs of differential subsidence, such as the vertical protrusion of more deeply buried sewer manholes from their spatial surroundings, often roads, leading not least to traffic hazards, although these phenomena may also be associated with other causes of subsidence (Rogers et al. 2006; on the human geographic classification of developments in New Orleans: Colten 2006). Drainage of peat soils can also be associated with significant subsidence, as much as 75% of its original thickness (Kolb and Saucier 1982). Extensive surface sealing (such as by buildings, roads, and plazas) in urban areas exacerbates soil subsidence; ultimately, the regenerative capacity of groundwater is diminished, and the water content of the soil decreases further (Rogers et al. 2006). The drainage of peat soils has a broad side effect with regard to the process of land loss—the oxidation of peat soils, which virtually 'vanish into thin air' (Rogers et al. 2006).

Another cause of subsidence is the increase in loading resulting from human activities. As shown, sediments deposited on the Louisiana coast, like peat soils, have a low degree of compaction. Overloading in this context is associated with subsidence. The nature of the loads can vary: They range from loading caused by levees (natural, but especially artificial) and those made by private individuals seeking to compensate for the subsidence around the foundations of their buildings to buildings themselves and technical infrastructures. These settlements caused by human loading do not only occur near the surface, but also extend into the deeper layers. This is suggested by the settlements of deep-foundation high-rise buildings in downtown New Orleans, and it is these high-rise structures that are associated with particularly large loads. Thus, the different magnitudes of loading are associated with different depths of settlement; whereas the twentieth-century subsidence under the central business district of New Orleans with its high-rise development was about 5 inches, the subsidence reaches only 2 inches under natural levees of the Mississippi River (Dokka 2011; Rogers et al. 2006). A graphic overview of the processes of land subsidence as an outcome of groundwater withdrawal, oxidation of peat soils, and loading can be found in Fig. 3.6.

Surface drainage is not the only form of material removal from sediments in south Louisiana that results in surface subsidence; there is also the removal of gas and fluids. For example, groundwater withdrawal was identified as a cause of surface subsidence as early as the 1960s. Focal points of this subsidence are noticeably found near deep withdrawal points for industrial use (Kazmann and Heath 1986). Another contribution to land subsidence is made by the extraction of crude oil and natural gas (see Fig. 3.7). The extraction of liquids and natural gas is associated with a reduction in supportive pressure. The pressure loss, in turn, often causes subsidence of the overlying strata (Barras et al. 2004; Dokka 2011; Olea and Coleman 2014; Rogers et al. 2006). Thus Dokka (2011) assumes a subsidence of local flood control structures and bridges by up to about 80 cm,

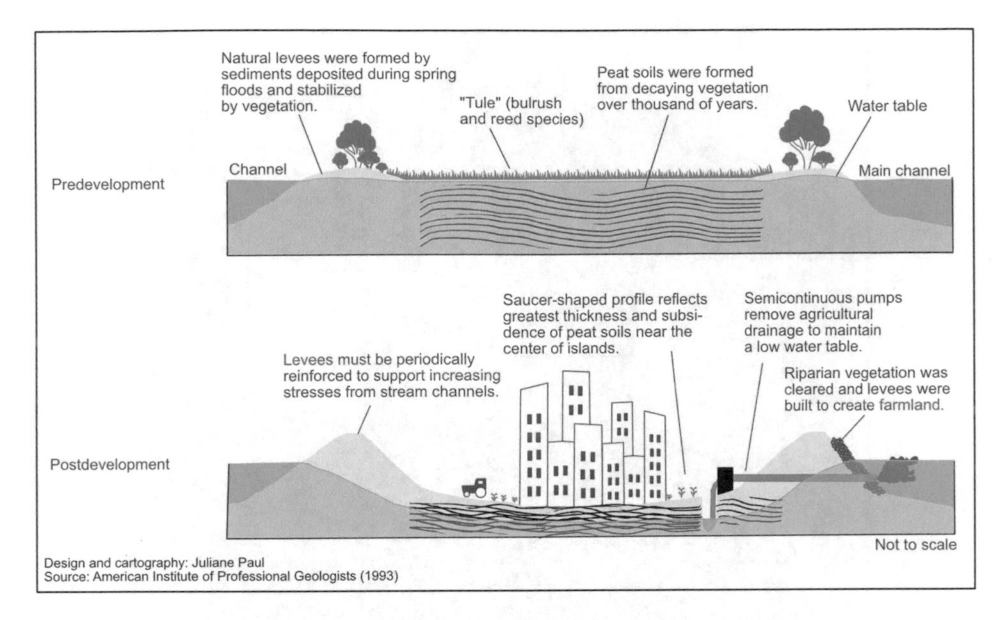

Fig. 3.6 Processes related to subsidence of the earth's surface on peat soils. (Heavily modified and added based on: American Institute of Professional Geologists 1993)

which could be attributed to groundwater extraction from aquifers at a depth of about 160 to 200 m in the urbanized areas of south Louisiana (especially New Orleans).

Another contributor to land subsidence is the dissolution of the salt domes already mentioned in the context of the formation of pressure ridges and fold belts. Salt domes form (diapirism) after the deposition of salt and subsequent uplift of surface sediments when salt is forced upward along zones of weakness and accumulates there. When water penetrates these salt domes, the salt is dissolved and discharged with the flowing water. As the salt (and sediment) is discharged seaward, the overlying material sinks, often creating exposures of surface water (Heinrich et al. 2020; Rogers et al. 2006). This dissolution of salt domes has been accelerated at some salt domes by human activities: In the context of the extraction of sulfur, water enters the salt domes and accelerates their dissolution or sets it in motion in the first place (Olea and Coleman 2014).

The wetlands of southern Louisiana are crisscrossed by thousands of miles of branching networks of pipelines and canals (see Figs. 3.8 and 3.9). These were constructed primarily for the purpose of transporting oil and gas, but also to supply machinery, spare parts, etc., and are associated with unintended side effects. As an example, wave action from boat traffic leads to increased erosion of unprotected shoreline areas, at the same time, saltwater infiltrating through the channels also leads to the death of autochthonous vegetation which is associated with the consequence of a greater risk of erosion of the now exposed shoreline areas (see Fig. 3.10). The same happens as a result of the escape of oil,

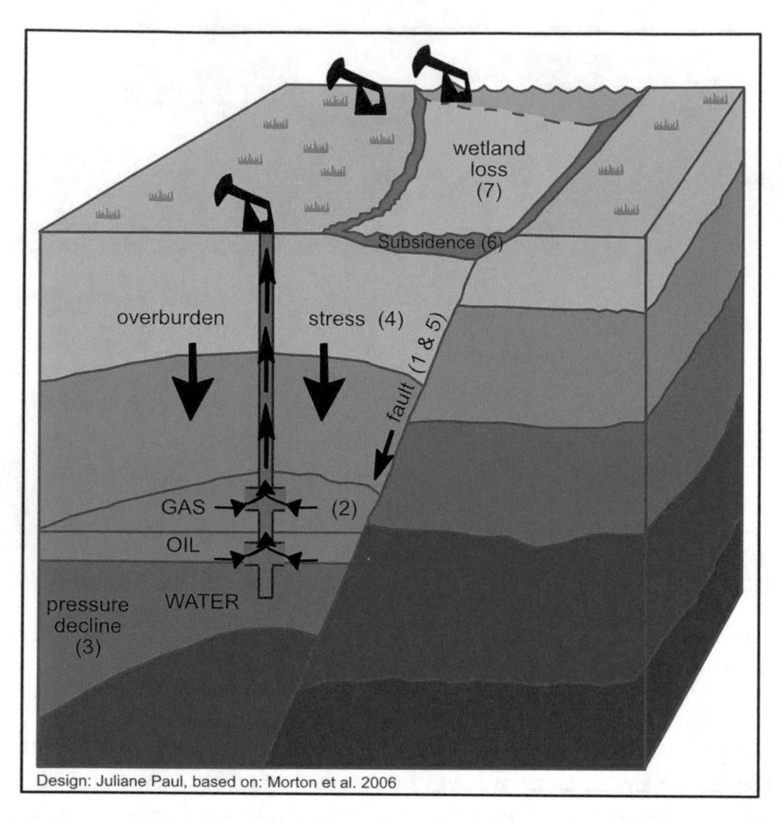

Fig. 3.7 Creation of wetland loss as a result of oil and gas production. A collection of oil and gas occurs along a fault (1). Extraction of oil, gas, and groundwater (2) results in subsidence pressure loss (3). The effective vertical stress of the overburden increases (4). This results in the compaction of the reservoir rock. This can lead to reactivation (5) of formerly active faults (1). Both compaction of the reservoir and surrounding strata and the sliding along fault surfaces can cause ground subsidence (6). If these compactions or fault-induced subsidence occur in wetlands, the wetlands are usually flooded and converted to open water (7). (Own presentation based on: Morton et al. 2006)

especially during its production and transfer, primarily as a result of accidents, wherein the oil spill as a result of the explosion of the Deepwater Horizon production platform is particularly significant (Barras et al. 2004; Bass and Turner 1997; Craig et al. 1979b; Gotham 2016; Hemmerling 2007; Hemmerling et al. 2020; Hester et al. 2016; Houck 2015; King 2017; Priest and Theriot 2013; Scaife et al. 1983; Shaffer 2016; Silliman et al. 2012; Theriot 2014).

Another unintended side effect of human activity contributing to land loss in Louisiana is the invasion of invasive species. For example, the decline of vegetation that protects against erosion is not solely from the construction of facilities to extract and transport oil and other commodities, but also from invasive species. For example, in the early 1900s an animal from Argentina similar to the beaver, the nutria, was introduced to enhance

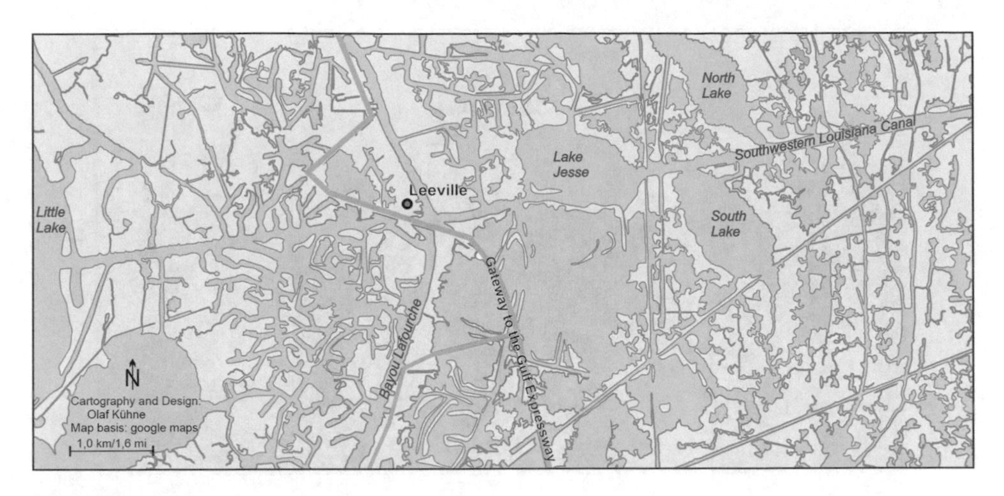

Fig. 3.8 An example of anthropogenic excessive impacts on Space 1 by humans through the construction of canals. (Own representation)

Fig. 3.9 The Old Intracoastal Waterway, second only to the New Intracoastal Waterway, as an example of the profound modification of Landscape 1 of southern Louisiana through the construction of canals. (Photo: Lara Koegst 2022)

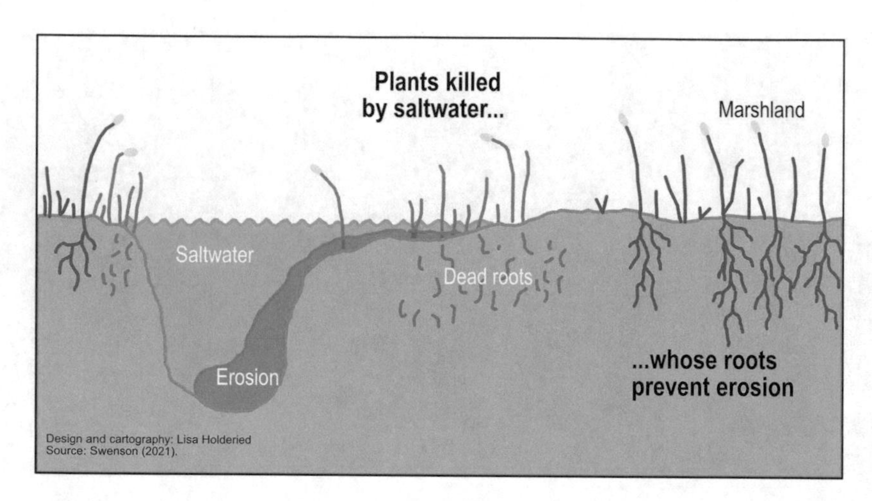

Fig. 3.10 Erosion in Louisiana streams as a result of saltwater intrusion, outflow from the petrochemical industry, and wave action from boat traffic, respectively. (Own representation based on: Swenson 2021)

fur production. Escaped specimens from fur farms were able to reproduce prolifically in the absence of natural enemies, increasing their population in southern Louisiana from a relative few to the millions existing today. They eat bare large areas of marsh grass and roots, leaving large areas exposed to erosion. Originally from China, the Roseau cane scale, an insect that feeds on sap from Roseau cane which can grow up to 10 ft tall, contributes to the death of this natural vegetation cover exposing marsh soil to erosion (Swenson 2021). Not only invasive species of animals contribute to land loss, but also (at least temporarily), invasive species of plants, such as the water hyacinth, which was distributed by an exhibitor in 1884 during the International Cotton Exposition in New Orleans. He distributed it as a gift, but it not only spread in the private garden ponds of each person who received the gift, but quickly spread to water bodies in Louisiana. Not only did it become a problem for navigation, but it also reduced the rate of runoff from the bayous during floods, which was countered with technical measures such as special mowing vessels or simple dredging, among other things. It also displaced native species, damaging local ecosystems (Wunderlich 1964; Yodis et al. 2016; see Fig. 3.11). In addition to extreme precipitation, storms—especially hurricanes—lead to land loss (Cahoon et al. 1995). For example, since 1879, the Louisiana coast has been hit by hurricanes, of at least Category 3, on average every 7.88 years (Turner et al. 2006). With the increase in anthropogenic climate change, hurricane events are projected to increase in number and severity (Mousavi et al. 2011). This means that the ability of the coast and its inhabitants to recover from each event is less and less likely. This recovery involves the deposition of material as a result of the tidal range, flooding as a result of heavy precipitation

Fig. 3.11 An example of a water body completely overgrown with floating plants; the St. Louis Bayou northeast of Isle de Jean Charles. (Photo: Lara Koegst 2022)

events, and also the replacement of coastal defenses or the restoration of buildings and infrastructure (Barras et al. 2004).

Another climate change-related process exacerbating land loss in Louisiana is sea level rise, although projections here are subject to a high degree of uncertainty, not least as a result of the potential for humans to influence greenhouse emissions (Kemp et al. 2011; Olea and Coleman 2014). Projections are for sea level rise on the northern coast of the Gulf of Mexico from 0.34 cm to 1.9 m by the end of the twenty-first century relative to that at the beginning of the century (Glick et al. 2013; Mousavi et al. 2011). According to these projections, a further loss of 2,188.97 km^2 (i.e., 9% of the wetland area calculated in 2007) to a possible loss of 5,875.27 km^2 (587.527 ha, 24% of the wetland area in 2007) is assumed for the aforementioned period (Glick et al. 2013; Kolker et al. 2011).

3.4 Coastal Land Loss in Louisiana: A Review of Spatial Extent, Temporality, Cultural-Natural Hybridity, and Human Mitigation and Adaptation Options

From a geologic/geotectonic perspective, the major contribution of regional incursion by the Gulf of Mexico is gathered from adjustment to sedimentary load in the form of lithospheric flexure, but also from normal faults dipping toward the basin (about 70%).

Compaction of the sedimentary material accounts for another 23%. The remaining processes generally occur localized, such as the extraction of groundwater, oil, and natural gas (Olea and Coleman 2014), but also the dissolution of salt domes—isostatic equalization forms a more long-term framework. In contrast, the subsidence processes of human origin, such as the consequences of the draining of marshes and swamps, the loading caused by buildings and technical infrastructures, and the reduction of sediment accumulation from the regulation of flowing waters, especially the Mississippi River, act in a very short time frame. Although inland land loss also occurs (such as along bayous or through subsidence resulting from the extraction of water, natural gas, and petroleum), loss along the coastline dominates (Britsch and Cunbar 1993; Britsch and Kemp, III 1991; Hester and Mendelssohn Irving A. 2000). A summary overview of the causes of land loss within their extent in Space 1, their temporal duration, their hybridity between the poles of natural processes and the imprinting by humans, and the latter's ability to influence them in terms of mitigation and adaptation comprises Table 3.1.

Table 3.1 Summary of causes of land loss in southern Louisiana. (Own compilation)

Process	Spatial extent	Time course	Culture Nature Hybridity	Mitigation/Adaptation
Glacial isostatic compensation	Continental	Long-term (decade thousands)	Nature	Adaptation only
Geosynclinal subsidence	Subcontinental	Long-term (decade thousands)	Nature	Adaptation only
Isostasy from water loading	Regional	Long-term	Nature	Adaptation only
Sediment reaction	Regional	Long-term	Nature	Adaptation only
Listric removals	Regional	Long-term	Nature	Adaptation only
Sediment loss from stream flow regulation.	Regional	Centuries	Anthropogenically induced	Mitigation only possible with severe unintended side effects, therefore adaptation
Drainage of organic soils	Multiple local/point	Centuries	Anthropogenically induced	Mitigation and adaptation
Dissolving of mineral substance	Multiple local/point	Centuries	Anthropogenically induced	Mitigation and adaptation
Compaction resulting from loading	Multiple local/point	Decades (increasingly since modern era (high-rise buildings))	Anthropogenically induced	Mitigation only possible with severe unintended side effects, therefore adaptation
Extraction of gases and liquids	Multiple local/point	Decades (Increasing since modern era)	Anthropogenically induced	Mitigation and adaptation

(continued)

Table 3.1 (continued)

Process	Spatial extent	Time course	Culture Nature Hybridity	Mitigation/Adaptation
Salt dome dissolution	Multiple local/point	Since the formation of salt domes, intensified decades (intensified since increased human intervention in surface formation)	Anthropogenically induced	Mitigation and adaptation
Canal and pipeline infrastructure	Multiple local/linear	Decades	Anthropogenically induced	Mitigation and adaptation
Increased erosion from the impact of invasive species on native vegetation	Regional	Decades (increasingly since modern era)	Anthropogenically induced	Minimize consequences, otherwise no more possibilities to influence
Impact of invasive plant species	Regional	Decades (increasingly since modern era)	Anthropogenically induced	Minimize consequences, otherwise no more possibilities to influence
Impact of storms and heavy precipitation events (especially hurricanes).	Regional	Always (intensified by anthropogenic climate change)	Natural, enhanced by anthropogenic processes	Mitigation with a long-time horizon and adaptation
Sea level rise	Global	Since the end of the last cold period (intensified by anthropogenic climate change)	Anthropogenically induced	Mitigation with a long-time horizon and adaptation

4 The Limits of Representation and Explicitness in Cartographic Representations—A Neopragmatic Redescription of the Louisiana Coast as a Hybrid Spatial Pastiche

In the foregoing, we have compiled the current state of research on coastal land loss, outlining, comparing, and categorizing the underlying processes. In doing so, processes of the natural-cultural hybrid Landscape 1, such as the isostatic compensation movements which are closer to the pole of nature, are augmented being partially superimposed by the interventions of Landscape 3 via Landscape 2 in Landscape 1 and in particular by the unintended side effects that occurred as a result of the interventions upon Landscape 1. The changes in Landscape 1 and their causes are subject to observation by Landscape 2 and are fed into the knowledge, interpretations, and valuations that constitute Landscape 3. As stated at the beginning of our book, the construction of landscape by humans occurs in different modes, that of the homeland (Mode A), that of common perception (Mode B), and that of 'expert special knowledge' (Mode C). We will focus on the Modes A and B later, turning first to a special construction of cognition in the Mode C, which enjoys special significance in dealing with land losses (Colten 2018; Khalil 2018)—that of cartographic representation. In this chapter, we return again to the fourth extreme mentioned in the introduction: the challenge, the complexity of capturing, and representing of the relevant phenomena. Here, however, not abstractly and metatheoretically as in Chap. 2 but concretely by means of cartographic representations.

In what follows, we undertake a threefold redescription. We begin by redescribing cartography in light of the theory of three spaces or three landscapes introduced in Sect. 2.1. We then demonstrate the limitations of a positivist approach to cartography that focuses on explicitness, and—aware of cartography as a twofold mitigation of complicatedness, complexity, and contingency—we propose a redescription of a cartography sensitive to hybridity. We then justify this ontologically by means of a redescription of coastal Louisiana as a spatial pastiche possessing varying degrees of hybridity.

O. Kühne and L. Koegst, *Land Loss in Louisiana*, RaumFragen: Stadt – Region – Landschaft, https://doi.org/10.1007/978-3-658-39889-7_4

4.1 Cartographic Representations—A Redescription in the Light of the Theory of Three Spaces

Applying the theory of three spaces, communication can also be traced by means of cartographic representation (Fig. 4.1; in the remainder, we follow in this: Kühne 2021d; Kühne, Edler, and Jenal 2022b). Central criteria for the creation of a cartographic representation lie in the assumptions of an intersubjective and an intertemporal relevance of what is represented, based on the social stock of knowledge, with Mode C access being particularly relevant here (in professional cartography). Cartographic representation can be understood as a hybrid of Space 1, 2, and 3, it is tangible (virtual reality and augmented reality also have a physical basis), it is individually conceived and even produced, and it is at the same time based on semi-societal conventions of what may be represented and in what way. At the same time, the cartographic representation—following the considerations of Niklas Luhmann (1987)—as a carrier of information and communication describes the synthesis of the three selections in a unity of information, communication, and understanding. Involved in communication are (in the simplest case) ALTER (the sender) and EGO (the receiver). ALTER has a socialization in the Mode C (such as the study of geography, geodesy, cartography, etc.). Thus, he has subject-specific conventions of observation (step 2), evaluation, and selection (step 3) of Landscape 1—in this case Landscape 1c. He translates these into a cartographic representation (here a map). In doing so, he resorts to the conventions of representation learned in the socialization of 'expert special knowledge' (Mode C). The map thus created as a hybrid of Landscape 1, 2, and 3, is in turn encountered by EGO on the basis of her/his Mode B socially learned conventions (cartographic common perception; step 5) and thusly observed (step 6). That is, EGO reads the map according to the learned interpretive conventions. The extent to which cartographic communication succeeds depends on whether and to what extent a relevant intersection of Mode B and C cartographic conventions exists (step 7).

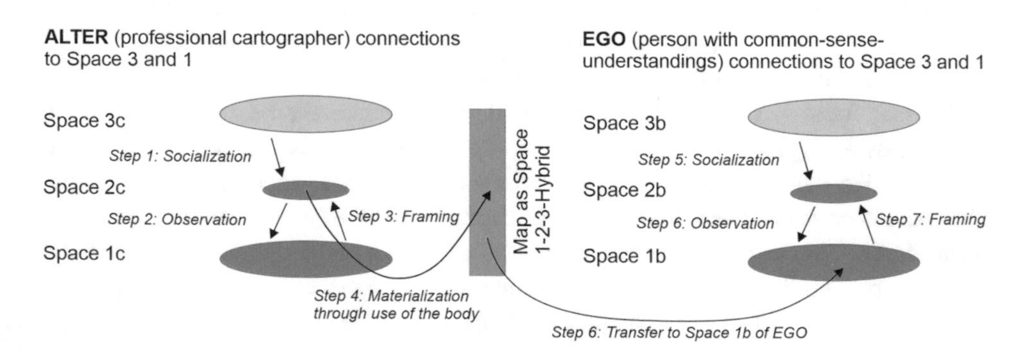

Fig. 4.1 The process of generating a map. (Modified after: Kühne 2021d)

In the following, we turn to the question of the construction of coastal land loss in the Mode C of cartographic representation in order to point here—quite exemplarily—to the challenges with the construction of explicitness and contingency reduction.

The sciences of cartography and geography grew out of the day-to-day worldwide challenge of finding one's way around the Earth—particularly beyond one's own typical perceptual horizons. The cartographic construction of the world, which went along with the improvement of finding one's way, was in turn connected with a double reduction of complexity and complicatedness: First, that of World 1 to Space 1 or to its special case Landscape 1, second, the cartographic representation on the basis of the reduction of complicatedness and complexity of World 1 to Space 1/Landscape 1. The development of standardized procedures of cartographic recording as well as cartographic representation ('signature language') meant the removal of geography/cartography from the Mode B and its transfer to the Mode C, i.e., the formation of a specialized discipline (Edler and Kühne 2022; Kühne 2021a). The transition into the Mode C was accompanied by standardizations, which can be understood as a reduction of contingency: In terms of science theory, a positivist world view is widespread, World 1 can be grasped with measuring, weighing, and counting, and a representation that is as true to scale as possible. The procedures and forms of representation are largely standardized (especially in topographic cartography), this standardization is based not least on the generation of clarity according to binary coding, such as "here is an object, there is not" or "here it is like this, there it is different" (Belina and Miggelbrink 2010; Kühne, Edler, and Jenal 2021b; Wardenga 2001a, 2001b, 2002, 2006). The effort connected with the cartographic registration of World 1, as well as the availability of spatial information, in turn, let this become a governmental task (Blackbourn 2007; Edler and Dickmann 2019). Both the positivist worldview and the power-bound nature of standardizations of cartographic representation, as well as the state, became the subject of critique by an emerging 'critical cartography' (among many: Crampton and Krygier 2005; Harley 1989; Kim 2015; Wood and Fels 1986). Their consequence lay ultimately in the efforts to transfer cartography from being Mode C bound to Mode B. This in turn reaches its limits with complex procedures (such as use of orthophotos) as well as objects (such as a changing coastal margin, as in Louisiana). A 'neopragmatic' cartography shares the critique of a reductionist worldview of 'traditional' cartography in Mode C, but arrives at a different consequence, namely the cartographic representation of contingency (Edler and Kühne 2022; Kühne 2021a; Kühne, Edler, and Jenal 2022a).

Such an understanding of postcritical cartography (Edler and Kühne 2022; Kühne 2021a) can be described as a redescription of cartography that does not paradigmatically reject both positivist and critical cartography, but strives to abolish their functional components. In doing so, it follows the approach of critical cartography insofar as to understand maps as social (and individual) constructions, constructions that have been experienced. These power relations are to be questioned in their effect on the enhancement of life chances, insofar as the approach does not follow the request to replace a

Mode C cartography with Mode B mapping. Accordingly, the results of 'traditional' positivist cartography are also not understood as a 'mapping' of 'reality', but as results of socially constitutive processes, which have to be examined with regard to their suitability towards maximizing life chances.

This is precisely the approach that becomes pertinent in the cartographic representation of land loss in Louisiana, as we will show below.

4.2 Of the Challenge to Draw a Clear Line: Cartographic Representations of the Coastline

The positivist framing is evident in the measurement of Louisiana's land loss: for example, Louisiana's net land loss during the 2004 to 2008 hurricane years (in addition to Katrina and Rita [2005], Ike and Gustav[2008]) is reported to be 327.9 miles^2 (849.5 km^2), which exceeds the net land loss occurring between 1978 to 2004 of 286.9 miles^2 (743.3 km^2; Barras 2009). The effects of Hurricanes Katrina and Rita would increase Louisiana land loss in 2005 to a record 118 miles^2 (305.6 km^2; Rogers et al. 2006). Barras calculates the storms added a net area gain of 217 miles^2 (562 km^2) of new standing water in coastal Louisiana (Barras 2007, 2009). Net area lost refers to the difference between land gains and land losses. Land gains result from the deposition of flotsam and jetsam along with the realignment of existing marsh areas displaced by storm surge.

Surveying land gains and losses relies chiefly on remote sensing methods, sensors that are not in direct contact with the earth's surface, to convert georeferenced data into two-dimensional and sometimes three-dimensional models (Weißmann et al. 2022). Thereby it is assumed—in the positivistic tradition of thought—that there is an effective accuracy between what is being depicted and the depiction of it (i.e., the depiction shows what is being depicted in such a way that it can be easily recognized). In the sense of a positivistic Mode C binary coding of water and land, which is oriented towards preciseness, this approach has its limitations. The methods of remote sensing, which are used in this context do not, or hardly allow, classifying of certain aquatic vegetation and driftwood with clarity between one and the other; also water level fluctuations resulting from normal tides and meteorological influences clearly limit the possibility of a clear-cut classification here, since an area overgrown by vegetation is only considered as mainland when the roots of the plants reach into the ground (Barras 2007, 2009; Couvillion et al. 2011). Regardless of the uncertainties presented, a detailed cartographic treatment of land loss in Louisiana is provided (here at: Couvillion et al. 2017)—adhering to the construct of dichotomous separation of sea and land: both the scale chosen (1:265,000) and the temporal resolution of the three-year increments manifest a high standard of detailed representation. The limits of modernist thinking in dichotomous precision (here in relation to the coastal area of Louisiana; Bauman 1990; Latour 1993; Welsch 2002) become clear here: the assumption inherent in the positivist worldview of the specific Mode C (geosciences) of that world (in this case Space 1)—that it can be unambiguously captured (and modeled on this basis) by measuring, weighing, and counting—is confronted with the challenge that the Space 1 under study not only eludes precise and stable assignment to the categories of land area and water area, but that the complexity of the processes causing coastal land loss (Chap. 3) makes it difficult to model future conditions. This is all the more true because the processes partially influence each other (prominent: petrochemical industry and climate change) and, opposingly, measures are taken for coastal protection, the effects and side effects of which often cannot be predicted exactly (Coastal Protection and Restoration Authority of Louisiana 2017; Hemmerling et al. 2022). The associated contingency of future developments (basically: Popper et al. 1994) in turn is (apparently) made less complex by the calculation of different scenarios (Couvillion et al. 2013). Everyday worldly danger is transformed into a calculable risk by applying the logics of Mode C (Bauman 2010; Beck 1986, 2006).

In line with the neopragmatic approach of our work, we do not want to deal solely with communications about world, space, and landscape in Mode C and their logics, but also with those of the other modes, principally their reciprocal effects. In the following, we deal accordingly with cartographic representations, which are cartographic representations being primarily addressed to persons who have no Mode C qualification in this respect (e.g. in cartography, geology, geography; Figs. 4.2, 4.3 and 4.4). These are related to a representation that does not follow the thinking in dichotomies (Fig. 4.5). The cartographic representations on which these maps are based were digitized for better comparability and subjected to a uniform signature language. In addition, a generalization to a comparable level and a representation responding to the question concerning the varying interpretation of land and water areas in different current cartographic representations of southern Louisiana was endeavored. With the goal of enhancing the understanding of the different representations of land–water distributions, all four maps were designed to represent the same roads, settlements, and linear water body structures. The selection was based on the criterion of distribution (according to the search behavior in Mode B; approx.: Linke 2020; Loda et al. 2020; Medyńska-Gulij 2012; statista 2019; Sullivan 2016; Yeo and Knox 2019), using the Nations Online maps (first hit of a Google search), the Google Maps map, and the Rand McNally map, as available physical maps. NASA's satellite image from May 12, 2022 was used to compare the current extent of the coastal area—on that day, the sky over Louisiana was unclouded and the atmosphere was barely clouded by haze, which facilitated the analysis of the satellite image (Löffler 1985; Schneider 2019; Walz 2001).

The comparison of the maps in Figs. 4.2, 4.3 and 4.4, especially in comparison to Fig. 4.5, shows clear differences in the shaping of the coastline between each, but also the representation of inland waters plus the naming of bays and lakes, which in the selected scaling of Google Maps (Fig. 4.4) and also the NASA image (Fig. 4.5) completely omit. The base map of Nations Online, as well as Rand McNally, classify significantly more areas as 'land' than the map of Google Maps, and especially the map derived from the

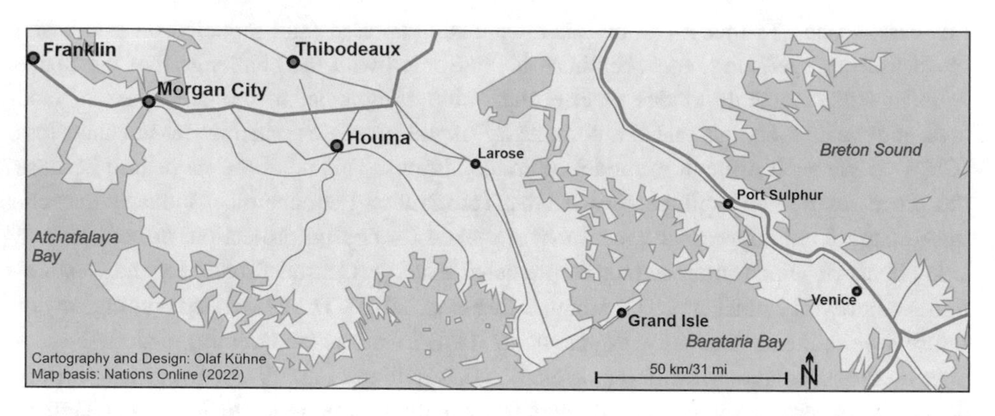

Fig. 4.2 Central coastal Louisiana based on Nations Online representation. (Own representation, map basis: Nations Online 2022; similarly published in: Kühne and Koegst 2022)

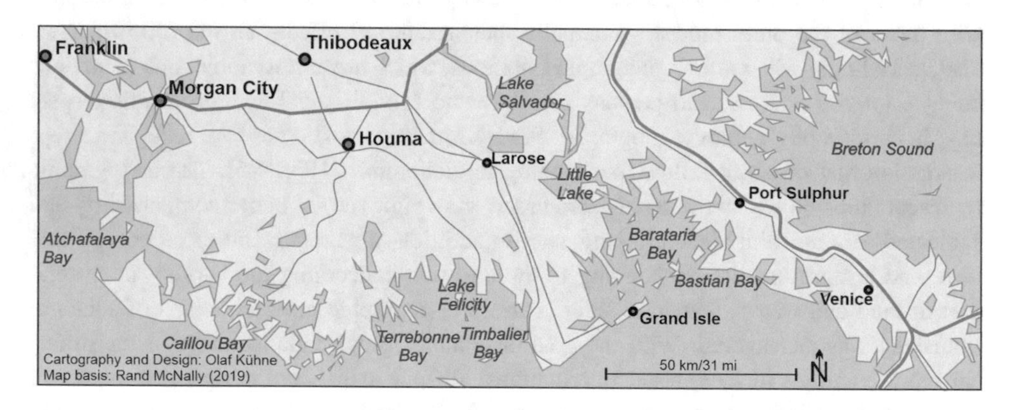

Fig. 4.3 Central coastal Louisiana based on Rand McNally's physical map representation. (Own representation, map basis: Rand McNally 2019; similarly published in: Kühne and Koegst 2022)

satellite image of NASA (Fig. 4.5). Noteworthy in both Google Maps and NASA is the discrepancy between the coastlines explicitly shown in the templates with the cartographically classified land areas (Google Maps) and the representations that can be clearly classified as land from the satellite image (although this may be influenced by tides, still this makes the need to precisely represent a hybrid space[1] even more significant). From Google Maps, and especially from NASA, in the western part of the map particularly, it is clear that the divergence between the depicted coastline and the area classified as 'land'

[1] Birch and Carney (2020) speak in this context of a "Coastal-Inland Continuum", which points in the direction of hybridity, while the choice of the word "continuum" suggests a constant transition, it does not have the manifold theoretical references that are connected with "hybridity". In this respect, we prefer the concept of "hybridity".

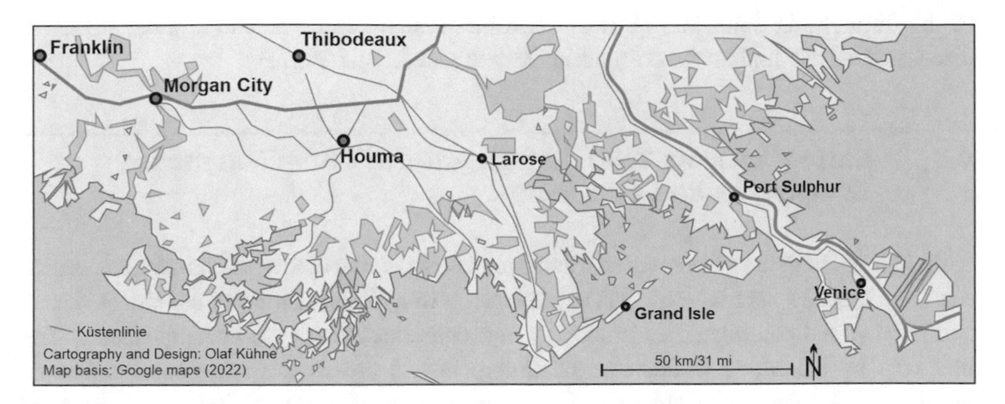

Fig. 4.4 The central Louisiana coastline based on a Google Maps rendering of the coast area. (Own representation, map basis: Google Maps 2022; similarly published in: Kühne and Koegst 2022). Of note here is the depiction of the coastline, which is not shown in other scales and is always shown slightly offset from the areas in the scale used here (which has been corrected in the map produced here)

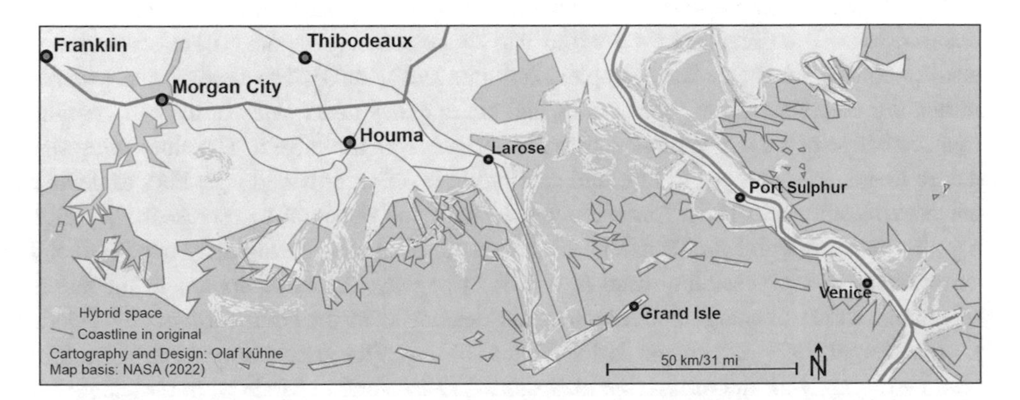

Fig. 4.5 The central Louisiana coast based on NASA's 12/5/2022 satellite image representation. (Own representation, map basis: NASA 2022; similarly published in: Kühne and Koegst 2022). The representation of the coastline contained in the satellite image was used in 'base map' manner, a colored line to depict hybrid spaces was added to the areas that could be precedingly classified as land or water areas, i.e., areas that could not be clearly classified here or had a small-scale alternation of land and water areas.

is strikingly large. Notably from the satellite image of NASA it is clear that between the defined coastline and those areas that can be undoubtedly classified as 'land' there exists, sometimes, several dozens of kilometers. It is in these areas that the loss of land is particularly evident. It is also remarkable that inland coastal areas are defined in Google Maps (indicated by orange triangles in Fig. 4.4). This can be understood as an indicator of the difficulty in delineating a clear coastline in southern Louisiana (a more intensive discussion of this topic can be found in: Kühne and Koegst 2022).

4.3 Cartography as Twofold Mitigator of Complicatedness, Complexity, and Contingency

Although not unique to the southern Louisiana coastline, the clarity of cartographic representation emerges here with distinct vehemence: a twofold mitigation of complicatedness, complexity, and contingency. Complexity and complicatedness are accomplished in the first step by the Mode C selection of structures and processes in the construction of Space 1c or Landscape 1c, the reduction of contingency is accomplished by the claim of superiority—blind to the socially defined selectivity of one's own landscape formation—of the Mode C landscape construction, following the positivist worldview, over Mode A and B constructions (a circumstance that 'critical cartography' rightly criticizes, because it limits the development of suitable approaches to 'world'; Popper 1963, 1992). Already on this level the effort to press the diversity of 'world' is predominant with its transitions and hybridizations into a scheme of unambiguity: climate classifications and natural spatial divisions. The division of the world into dichotomously separated landscape zones testifies to this effort (see for example: Zierhofer 1999, 2003). The second step towards mitigating complicatedness, complexity, and contingency takes place in the cartographic transformation of the explicitness apparently derived from the subject. The elucidating signature language knows point, line, and area symbols (often provided with lines to delimit the areas from each other). Surface symbols have the potential to represent hybridity, e.g., by means of color gradients, but neither in classical atlas cartography nor in official cartography is this possibility used in a distinctive way; the same applies to the forms of representation of margins and boundaries developed in thematic cartography (Mode C) (Arnberger 1993; Dickmann 2018; Kühne 2022a). The great importance of straight atlas cartography for the production of Space 2/Landscape 2 in Mode B, in turn, signifies the socialization of a modernistic and dichotomous worldview oriented to the projection of boundaries as an essential process of transformation from World 1 to Space 1/Landscape 1 (Kühne 2018c), although World 1 is much more strongly characterized by edges and transitional boundaries (Ipsen 2006). Thus, a recursively reinforcing feedback process takes place between Mode B and C, in that the standards shaped in Mode C (Fontaine 2019, 2020b, 2021) especially in atlas cartography (similarly: street mapping), are socialized in Mode B (step 5 in Fig. 4.1), which in turn are updated by the same patterns of constructing explicitness, going back to modernist thinking, through ever new representations of maps in Mode C. In this respect, using the Louisiana land loss example here, there is also the Mode B expectation of the dichotomous representation of the advance of water by means of distinct water and land boundaries, not least to illustrate the political explosiveness of the issue (Colten 2018, 2021b; Warnke 1992). The political impact of maps in the context of land loss in Louisiana is demonstrated largely by "The Green Dot Map" as published in the Times-Picayune on January 11, 2006. The map depicted New Orleans neighborhoods damaged by flooding as a result of Katrina. A moratorium on building permits was proposed for these areas, which temporarily excluded them from rebuilding, which in turn mobilized those New Orleanians that had subsequently spread across the country to massive protests, preventing further encroachment on the settlement pattern (discussed in detail: Lamb 2020).

The example of cartographic representation shows that while on an ontological level the suggestion of a clear separation of land and water areas does not lend itself to a suitable description of the contingency of the specific Space 1, conversely, political efficacy can certainly preserve life chances that otherwise could not have been preserved. In this respect, even this example shows the functional potential of a positivist cartographic tradition.

4.4 Of Hybridizations and Pastiches and the Cartographic Treatment of Them—A Re-description of Coastal Louisiana

It is precisely the areas between (formerly) defined coastline and areas that can be clearly classified as 'land' that we initially understand as hybrid space—including temporarily flooded areas, the mobility of marshlands, the dynamics of erosion and deposition—among the long-term trend of the processes of land loss outlined in the previous section. A conceptualization of the representation can be found in Fig. 4.5. The necessity of thinking in terms of hybridity, of saying goodbye to singularity, is particularly evident in Fig. 4.3 in the designation 'Little Lake'. Already in this map it is clear that the water area still classified as 'Lake' has been absorbed into a larger hybrid bay or cove area. This becomes even clearer in Figs. 4.4 and 4.5 where it is hardly possible to identify a distinct unit 'Lake' in the hybrid area between land and sea.

This challenge is not fundamentally new, even the early cartographers had to face it: "Rather than impressive topography, the coast is mostly a vast low-lying marsh where it is difficult to distinguish where water ends and land begins" (Colten 2018, p. 1). Accordingly, the cartographic representation of Louisiana's coastal zone was also vague in the early days of cartographic mapping of southern Louisiana. The subordination to binary thinking, combined with the striving for explicitness as well as a unifying (national, sometimes international) language of symbology, including outside of topographic cartography (Arnberger 1993; Dickmann 2018; Field 2018; Imhof 1972), made individual representations of hybrid spaces seem difficult to achieve. However, even if the challenge is not new, it acquires a new timeliness relative to the processes of land loss presented in the previous chapter to illustrate the different processes of land loss. The processes surrounding land loss illustrate the dynamics of shifting portions of area classified as 'land' to hybrid spaces, from hybrid spaces to areas of water. This trend can be slowed to some extent with coastal protection measures, creating a spatial pastiche (Hoesterey 2001; Kühne, Jenal, and Koegst 2020) of areas possessing varying degrees of land and water hybridity. The term pastiche can be used to describe spaces whose segments are characterized by hybridity and ambiguity. In this context, the concept of pastiche is able to demarcate and overtake the polarity of distinction and indistinction, because "pastiche does not simply mean de-differentiation, but presupposes difference formation, to then lead to hybrid intersections, recombinations, reintegrations" (Vester 1993, p. 29).

The Louisiana coastal zone can thus be described at the level of Space 1 as a pastiche of overlapping and differing structures and processes of varying intensity. Particularly evident in the foregoing was the component of varying degrees of water-land hybridity. However, the spatial pastiche is also characterized by varying degrees of nature-culture hybridity; after all, the coastal area is overlain by human activities to varying degrees of intensity. Different degrees of ruralness and urban development are also found in this space. In this way, a pastiche is created that is composed of segments of varying hybridities and degrees of hybridity.

Drawing on Richard Rorty's terminology, the twofold mitigation of complicatedness, complexity, and contingency outlined in the foregoing, in conjunction with the comments on hybridizations and spatial pastiches, demonstrates the extent to which land loss should be described in terms of invention rather than discovery. Not insofar as land loss did not occur at the level of World 1, but insofar as its transformation into Space 1/Landscape 1 is accomplished in Mode B and C, namely in the interpretive and representational patterns of Space 3b/Landscape 3b and Space 3c/Landscape 3c, respectively. While the common perception of Mode B still shows a certain coherence (although this also increasingly dissolves as a result of the differentiation of the information sources; see for example Kaußen 2018; Kühne, Koegst et al. 2021; Nagle 2017; Wagner 2019), the construction of space and landscape in the Mode C is highly nuanced: The loss of land in Louisiana in World 1 also shapes, very differently between each, the views of the historical geographer, the hydraulic engineer, the climate activist, the spatial planner, the commercial fisherman, the manager of an oil company, etc., and also shapes differently those of the long-established resident or the vacation home owner. This is true not only for the norms applied to Landscape 1, but also for the syntheses of World 1 to Landscape 1 that loop in feedback with them, a topic we will address in the following section.

Land Loss and Life Chances 5

Having dealt in the previous chapters with the extremes one, two, and four as presented in the introduction, that is, with the natural or human processes related to land loss, as well as with the scientific (or scientific and theoretical) issues of access and representations of the complex processes and structures related to it, we now turn to extremes three and five, that is, to the social issues related to land loss, as well as to the political responses to it. In this context it becomes clear that the chosen theoretical framework needs enhancement in order to better express the complex relations between the cultural, social, and economic aspects—this enhancement of the theoretical framework is done by adapting Pierre Bourdieu's theory of symbolic capital (Bourdieu 1984, 1985, 1989, 2002). In the second part of the chapter, we reflect on the suitability of this theoretical approach as well as the normative framework of our work, the 'life chances' approach of Ralf Dahrendorf. For both theoretical approaches, we conclude that they need to be augmented to include spatial and landscape aspects in order to understand the complex processes of extremes in Louisiana. Accordingly, we elaborate proposals for theoretical enhancements or differentiations in this regard.

5.1 Social Issues of Land Loss

Since the beginning of its settlement, coastal Louisiana has been characterized by an extreme polarization of opportunities and hazards. On the one hand, it offers areas for primary economic uses, such as fishing, oyster farming, hunting, trapping, sections for agriculture, and with oil a commodity that had become the fuel of economic prosperity during the course of the twentieth century. On the other hand, human life is threatened by floods and hurricanes. In recent years and decades, this relationship has shifted strongly in the direction of the hazards as a result of the developments described in Chap. 3 on hazards and will, in all probability, continue to develop in this direction—especially against

O. Kühne and L. Koegst, *Land Loss in Louisiana*, RaumFragen: Stadt – Region – Landschaft, https://doi.org/10.1007/978-3-658-39889-7_5

the background of progressive anthropogenic climate change (Crepelle 2018; Hemmerling 2007; Hemmerling et al. 2020; Trepanier and Scheitlin 2014). Impacted by the intensification of the risks is a population that was forced to settle along the bayous as a result of displacement (especially in the eighteenth century by Arcadian French from the Atlantic provinces of today's Canada as well as indigenous peoples) and has since developed an economic and lifestyle adapted to the local ecological conditions (Colten 2015; Colten et al. 2018; Colten and Giancarlo 2011; Dajko 2020; Jessee 2020; Maldonado 2015; 2019; Peterson and Maldonado 2006). As a result of the rising sea level exacerbated by the subsiding land areas, the possibilities of adaptive options to the rapidly changing conditions have been exhausted in many places. This concerns the possibilities of individual adaptation (for example, the construction of additional homes), community adaptation (for example, evacuation plans and emergency shelters), as well as state measures of coastal protection (such as the construction of breakwaters or ever newer levees; Boesch 2020). In line with the exhaustion of available resilience reserves, resettlement becomes inevitable in many cases (Baum 2021; King 2017). This resettlement is not evenly distributed according to the demographical age structure, since the outflow is predominantly the younger people, while older members of the population persistently remain loyal to endangered residential locations. Additionally, migration increases after catastrophic events (especially severe hurricanes; Colten et al. 2018; Jessee 2020). Long-time resident populations with low endowments of symbolic capital (in Bourdieu's (1984) sense of the term and correspondingly fewer options) are particularly vulnerable, as they often cannot afford to upgrade their flood-affected homes, flood insurance is also often too expensive, and moving inland is likewise hampered by a lack of economic capital and, in the destination region, social capital. For those with coastal jobs (such as in the fishing or oil industries) who can afford to relocate, commute times increase greatly (Jessee 2020; Laska et al. 2015). However, out-migration is also countered by specific in-migration: Since the 1970s, wealthy white "weekenders" have been building or acquiring a camp[1] on the coast to take advantage of the local Landscape 1 for recreational activities, combined with patterns of gentrification trends. A development strongly encouraged by municipalities as a result of hopes for higher tax revenues than from resident populations with their often low market value homes (Jessee 2020; Peterson and Maldonado 2006; Solet 2006). The camps of the weekend users are mostly clearly distinguished from those of the permanent residents of the coastal areas: they are usually built according to the flood adaptation regulations in force at the time and are thus erected on higher stilts (Solet 2006), so here the social indicator dimension of the 'appropriated physical Landscape 1' emerges clearly (Kühne 2006a).

Although coastal protection has recently shifted away from a focus on engineered flood protection to one of wildlife protection and wetland restoration, the needs as well

[1] On the Louisiana coast, camp refers to a house of smaller than average size built of wood, often on stilts. Nonetheless, the buildings are often used as primary residences and thus can be understood as the equivalent of a classic house.

as knowledge of local conditions of affected residents have long been largely disregarded in planning (Colten 2017). Only in the recent past—also against the background of the worsening situation of land loss in conjunction with the growing criticism of the lack of participation of the local population—has there been an increased consideration of their perspectives (Hemmerling et al. 2022). This is not the first change of strategy in the public handling of flood hazards; responding to the disastrous impacts of the Great Mississippi River Flood of 1927, the Flood Control Act of 1928 shifted the policy away from the construction and reinforcement of levees towards engineering massive dams and floodwalls, as well as to control structures and spillways (Boesch 2020). The increasing mechanization of flood control was accompanied by a shift of responsibility to higher levels (Boesch 2020; Colten 2021b). This in turn was connected with the dominance of the perspective of experts over the local population (but also the administration) as mentioned above. These large-scale water management facilities are still in use today, for example, where high water levels in the lower Mississippi River are still discharged into Lake Pontchartrain by opening the Bonnet Carré Spillway or across the Atchafalaya Basin via the West Atchafalaya or Morganza Control Structure (Boesch 2020).

In the context of coastal protection, the most comprehensive plan to date is the 2017 update to the state's 50-year Coastal Master Plan. Unanimously approved by the Louisiana Coastal Protection and Restoration Authority (CPRA) in April 2017, the update represents a significant expansion of the original 2007 master plan and its 2012 update (a fourth edition of the master plan will be published in spring 2023). A total of 124 projects are expected to preserve or reclaim 800 miles^2 of land, with the goal of avoiding $150 billion in flood damages over the next 50 years (Hemmerling et al. 2020). The Comprehensive Master Plan identifies numerous settlements where it will be impossible to build within the next 50 years as a result of rising water levels, even without the impact of hurricanes, these are (from east to west; Clipp et al. 2017):

- St. Bernard Parish: Delacroix,
- Plaquemines Parish: Venice,
- Jefferson Parish: Grand Isle, Lafitte/Crown Point/Barataria,
- St. Charles Parish: Paradis,
- Lafourche Parish: Kraemer, Leeville,
- Terrebonne Parish: Dulac, Cocodrie, Isle de Jean Charles, Lower Pointe-aux-Chênes.

This generated criticism in that it did not present a detailed planning program for the buyout purchases of properties and, simultaneously, the plan did not inform the affected residents they resided within impacted portions of the plan boundaries (Jessee 2020; Wendland 2018).

The challenge of relocation is not solely a technical one: "The displacement and transition of population, business, industry, and social functions from the coast is sometimes a hope to get to higher, drier, and safer places" (Peterson 2020, p. 191). At the same time, social structures are torn apart over decades or at least forced to adapt to new contexts, the loss of the familiar Landscape 1a increases insecurity, economic clusters are forced to adapt to new conditions, etc. The adaptation effort is not limited to the coastal areas. The adjustment effort is not limited to the migrating population, but also to those already settled in the destination regions. They, too, have to face the changed situation, also with regard to Landscape 1. But resettlement can also be seen as an opportunity: "Geographical population shifts can be an opportunity for the state to address direct social, health, and human dimension issues while in the midst of coastal land loss and forced displacement" (Peterson 2020, p. 209).

The extremes of Landscape 1 (and also 2 and 3) in southern Louisiana involving natural developments and human interventions in Landscape 1, concomitant social developments, and political responses, presented and revisited in the introduction, is also associated, as shown, with significant changes in the Mode C construction of landscape. This change in Mode C construction, in turn, has implications for the treatment of Landscape 1. For example, the Mode C notion of the controllability of natural processes contributed to the construction of extensive engineering facilities to regulate water bodies. The Mode C valuation of marsh and marshland as not being (economically) useless residual land made this Landscape 1 the object of extensive physical revision, such as through canal construction. In Mode C self-awareness, the resident population with their Mode A and B interpretations and knowledge of Landscape 1 was understood more as an object of planning, not as a resource of knowledge about actual Landscapes 1, or even (equal) partners with legitimate claims. These Mode C constructs have now given way to new interpretations, valuations, and categorizations: Confidence in the technical controllability of processes located at the pole of nature in cultural-natural hybridity, and also of the inherent dynamic side-effects of human interventions, is fading in the face of progressive land loss (and selective land gain). Marshland is increasingly understood not as an (economically) worthless residual category, but as a Landscape 1 having value as coastal protection. Also, an appreciation of Mode A and B perspectives is increasingly asserting itself in the Mode C consciousness of planning processes.

This change of Mode C perspectives, however, is not only an outcome of the realization that 'classical' paradigms cannot (any longer) sufficiently interpret the relations of World 1, but also World 3 and partly World 2, alternately being unable to (any longer) provide suitable suggestions in dealing with them. The Mode C logic uses 'world' (independent of 1, 2 or 3) as a medium for the development of atypical interpretations (Edler and Kühne 2022) to generate—should the deviance prove to be functional—scientific progress, optionally also to generate by this behavior of distinction an enhanced intradisciplinary reputation (see among many: Feyerabend 2010 [1975]; Kaiser and Maasen 2010; Kühne 2008; Laudan 1977; Weingart 2015). This in turn has the consequence that the above described changes of the Mode C construction of the world will again be the starting point of future Mode C deviations (Kühne and Berr 2021). After all—going back to the classical saying of Karl Popper—the future is open (Popper et al. 1994).

5.2 Reduction of Life Chances from Loss of Specific Symbolic Capital Via Land Loss—An Extension of the Ligature Concept and Specification of the Theory of Symbolic Capital

Rooted in what has been expressed about the social implications of the land loss along the Louisiana coast, we want to examine and, based on this examination, specify or extend this central theoretical framework of our investigation with the loss of Space 1 and thus the most extensive deprivation of the basis of human existence: the theory of symbolic capital by Pierre Bourdieu (1984, 1985, 1989) and the life chances concept (especially its part of ligatures) by Ralf Dahrendorf (1979, 2002).

The homeland ligature (in relation to Landscape 1a) highlights the ambivalence of ligatures: on the one hand, it has provided a physical substrate for bonds and specialized knowledge for the use of hybrid space for centuries; on the other hand, the foundations of Space 1 were so significantly altered by the loss of land, that this ligature robbed more and more options of livelihood, so that only relocation remained. This particularly affects segments of the population that have an overall comparatively lower endowment of symbolic capital of the long-established (beyond that of a social and cultural capital valid in the local environment) is associated with a diminishment of endowment options. If this local social capital (the cultural capital directed at Landscape 1 is already devalued by the loss of land) is destroyed by relocation measures, this means a massive loss of life chances. Not only options are thus taken away, but also implicit and internally rooted ligatures of orientation in a specific Landscape 1a become meaningless. The people concerned are confronted with new externally directed (often moral) ligatures. In this respect, the effort to carry out relocations of communities appears—against the background of the understanding of life chances developed here—to be an important opportunity to destroy fewer options and fewer meaning-giving ligatures than would have been feasible in the case of a forced individual relocation.

On a meta-level, these remarks also demonstrate that Pierre Bourdieu's concept of symbolic capital (Bourdieu 1984, 1985, 1989) requires elaboration in the context examined here. From the explanations above it becomes clear that symbolic capital can be subject to spatial differentiation. That is, it is neither universally nor socially valid uniformly, but can be sub-societal, in this case regional or specified. The living and economic conditions adapted to a specific Space 1 with very specific living conditions and developed in a relatively isolated way produce a kind of social capital as well as incorporated cultural capital, which is vital for survival in this specific Space 1, but (largely) devalued in other places of living and working. Conversely, the inscriptions of universal/total social access in Mode C (specific incorporated cultural capital institutionalized through educational titles) in Space 1c is shown to influence its specific natural-cultural hybrid development dynamics in such a way that adaptation measures by the local population are no longer possible and relocation remains the last remaining option.

The specific conditions in southern Louisiana with their loss of Space 1 and thus the most far-reaching deprivation of the foundation of existence for people, thus means a test for social theories. Not just Pierre Bourdieu's theory of symbolic capital needs revision here if the conditions are to be subjected to an appropriate theoretical framing. The same applies to the approach to life chances, which is central to our work, and in particular to the understanding of ligatures, which has already been expanded upon in Sect. 2.3. The ligature concept—as stated in the mentioned section—is, so far, limited to the relationships between World 2 and World 3, where it is especially directed at processes between person and society. In view of the dominant influence that itself is radically evolving in relation with a radically emerging Space 1, an integration of World 1, respectively Space 1 and Landscape 1, into the ligature concept seems to make sense in view of the impact of World 1 on the physical human presence in this space, in order to be able to integrate into the theoretical framework not only the cognitive integration of the individual (World 2) into the social and cultural World 3, but also that which is physically integrated into World 1 with its dependencies and conditionalities. In the following, we will thus deal with the repercussions of World 1 on World 2 and vice versa via mediation of the human being.

Humans are physical beings. Thus, they are subject to the inherent laws of World 1, but they are also able to modify their feedback with the rest of World 1 to a small extent. We conceptualize this regimentation of World 2 by World 1 as corporeal ligatures. These corporeal ligatures can be extended by technical measures (as though clothing by an extension of our limbs, as a means of conveyance), corporeal ligatures can thus be diminished, options expanded. Regardless of this, there remains a considerable residue of corporeal ligatures that humans physically cannot escape (such as the limits of the body's physical resilience). But also, the possibility of using the technical extensions and measures for the reduction of the number and effect of corporeal ligatures is again regulated by the dependence of World 1 upon World 3 (for example a car can be used only if there is also money available for its operation). Again, these are ligatures that affect World 2, these we call corporeally mediated social ligatures. Layered self-referentially in terms of the neopragmatism we advocate, this extension allows for an intensified conceptual incorporation of a phenomenological reference through a greater consideration of the corporeality of human beings (this, in turn, will become relevant in Chap. 7).

In the context of our study area, this extension means: the impacts of hurricanes and but also of the processes of coastal land loss on World 1 function as corporeal ligatures on World 2. Adaptation measures to land loss, such as building raised houses on stilts, building levees, or devising evacuation plans, represent attempts to reduce corporeal ligature liability as well as in number and intensity of impact. As introduced, these adaptation measures are not free of unintended side consequences in their connections in World 3: Exclusion, from the mainland area protected by a levee, results in physically imposed social ligatures, as in the case of Grand Isle or Isle de Jean Charles, among others. This, in turn, has a very limiting effect on the options of the resident populations until they are forfeited.

6 The Loss of Coastal Land in Mass Media Representation

Mass media have a central role in the social construction of world in general: space and landscape in particular. Notably, mass media communication provides a medium of connecting extremes such as those mentioned in the introduction; one and two (natural processes and human influences), as well as four and five (the social and the political reactions to them). Sometimes, however, Mode C interpretations are resorted to, thus integrating 'extreme four'. As a result of the ability of (mass) media communication to address society in its entirety and to bring all social subsystems into resonance (Luhmann 1996), it also seems to be a good venue for a horizontal geographic analysis examination of this construction of world, space, and landscape. To this end, we will first look at the function of mass media in society before we address a review of the state of research on the mass media representation of coastal land loss in Louisiana. We will supplement this overview with our own examination of communication involving a YouTube video about the impact of Hurricane Ida in August 2021. We will conclude this chapter by linking the empirical findings back to the theoretical framework.

6.1 The Function of the Mass Media

Mass media have played a significant role in generating World 3 and 2, as Burgess (1985, p. 194) states, "The mass media play a profoundly significant role in the appropriation and interpretation of the meanings of social reality. They have the capability to shape conceptions of our physical, economic, political, and social environments". Not only upon the production of World 2 and 3, but also a significant impact in relation to the production of Space 2 and 3 as well as Landscape 2 and 3 (among others, Escher 2006; Escher and Zimmermann 2001; Kühne, Koegst et al. 2021; Luhmann 1996; Ziemann 2012), inasmuch as it seems necessary to incorporate modes and processes of the media

O. Kühne and L. Koegst, *Land Loss in Louisiana*, RaumFragen: Stadt – Region – Landschaft, https://doi.org/10.1007/978-3-658-39889-7_6

construction of Louisiana's coastal land loss into a neopragmatic horizontal geography, accordingly, the goal of this study is to trace coastal land loss in its complex relations between Space/Landscape 1, 2, and 3. Exploring the media construction of space and landscape takes many forms between levels 1, 2, and 3, for example, it "asks how media and media content produce reality, shape education, make it experiential, and how (mass) media produce social and spatial order" (Harendt 2019, p. 89; see also Aitken and Craine 2015; Burgess and Gold 1985; Ziemann 2012; Zonn 1990). In this context, social science research on the relationship between space and society has differentiated in recent years (cf. i.a. Harendt 2019):

1. The media are examined in terms of their function in relation to space and society (see Döring and Thielmann 2009; Werlen 2008).
2. The media are considered visual geographies (among others: Dodge et al. 2008; Fröhlich 2007; Linke 2019; Schlottmann and Miggelbrink 2009, 2015; Zimmermann 2007).
3. The media are examined in context of the connections between geographies and technology (Felgenhauer 2011; Kitchin and Dodge 2011; Thielmann 2007).
4. The media are considered in terms of 'geomediality' and -visualization (Edler et al. 2018; Edler, Keil et al. 2019; Gryl et al. 2013; Jekel et al. 2012; provide an overview Bork-Hüffer et al. 2021).

In this context, mass media are not able to convey 'reality' or even objectivity; they provide mediatization—often artificially produced images of realities through "selection, perspective, coloring, lighting, language, and commentary" (Rüthers 1999, p. 14). Media thus advance to "reality devices" (Fröhlich 2007, p. 98) and "worldview generators" (Harendt 2019, p. 89). In terms of content, media-related landscape research deals with the representation and symbolic charging of landscape in the various media, such as websites and newspaper articles available on the Internet (Baum 2021; Berr et al. 2022; Jenal, Endreß et al. 2021; Weber et al. 2017), films (Escher and Zimmermann 2001; Lefebvre 2006; Lukinbeal 2012; Zimmermann 2019), in computer games (Fontaine 2017, 2020a; Kühne 2022b; Kühne, Jenal, and Edler 2020), of photographs presented on the Internet (Dunkel 2015; Kaußen 2018; Linke 2017; Loda et al. 2020; Wartmann and Mackaness 2020), as well as videos distributed on the internet (Kühne 2012a, 2020b; Kühne and Weber 2015; und Kühne and Schönwald 2015). In recent years, the Internet in particular has become an essential medium for the social construction of landscape, especially for the actualization of stereotypes of Mode B (Bernstein et al. 2019; Jenal 2019; Kühne 2018d). As a result of the development of Web 2.0 and social media, mass media communication has fundamentally changed in the process from a far-reaching distinction between institutions who provided viewers with information unidirectionally, such as private sector or public sector operated transmitters mostly serving specialized professional persons into the emerging structure in which the actors (can) take on both functions, creating the now frequently mentioned 'prosumer'. In addition to the development of multidirectional communication, the speed of mass media communication has also increased significantly during the onset of these developments (Althoff et al. 2016; Hensel et al. 2013; Kühne 2020b; Lemos 2010; Purcell 2018; Wagner 2019). In this respect, the study of Internet communication for the social construction of landscape in general, that of coastal land loss in Louisiana in particular, becomes a relevant topic (not only) for our work.

6.2 The Presence of Threats to the Louisiana Coast in Mass Media—An Overview of the Current State of Research

The Isle de Jean Charles which is, or was, mostly inhabited by Biloxi-Chitimacha-Choctaw Indians, has a distinct media presence regarding land loss and its consequences. This largely stems from the fact that this was the first state-organized relocation effort of coastal residents, and because the island's inhabitants are considered America's first 'climate refugees' (Baum 2021). Located in Terrebonne Parish, the island was settled in the context of the Indian Removal Era by fleeing Native Americans in the 1840s. The descendants of these settlers predominantly lived at subsistence level by fishing, hunting, and extensive agriculture with livestock (Colten et al. 2018; Maldonado 2019), interwoven with little reconfiguration of Landscape 1. However, their way of life was successively undermined by allochthonous reconfigurations of this Landscape 1. This refers not only to the complex developments of coastal land loss (Chap. 3), but also to the changes in the ecosystem resulting from saltwater intrusions that damaged those fish and crustacean stocks on which the island's inhabitants partially depended (Colten et al. 2018; Gotham 2016). Presently, the condition is different: "Today, adults longingly remember shrimping and crabbing with their families in the bayou that ran the length of the island—a sanctuary in an area once considered 'uninhabitable swampland' by government officials" (Jessee 2020, p. 148). From this, not only the rapid change of Landscape 1 in the area around the island becomes clear but also the fundamental difference of Mode A and B: Landscape 1a is the familiar area, which also provides the basis for one's own nutrition, as Landscape 1c it is a remnant area falling out of the economic, as well as political, rationale. These changes of Landscape 1 due to the loss of land is thus "not just physical but impact peoples' sense of place, belonging, cultural identity, and local knowledge" (Maldonado 2015, p. 205).

Isle de Jean Charles is a particularly dramatic example of coastal land loss: while its area was 2400 hectares in 1955, it had shrunk to 320 hectares by the mid-2010s (King 2017, p. 306). The particularly extensive land loss of Isle des Jean Charles has some specifics that go beyond the general processes of land loss: For example, the islands locaition outside the protection of the 98-mile Morganza-to-the-Gulf Hurricane Protection System levee, constructed in 1998 primarily to protect those parts of the country farther north (Schleifstein 2021; Fig. 6.1). The offshore barrier islands are also subject to increased erosion as a result of intensified hurricanes (Jessee 2020; cf: Fearnley 2009). This is substantially because of the prioritization of the protection of larger settlements, which ignores smaller settlements. The fact that indigenous communities in particular live in smaller settlements raises the suspicion of the persistence of colonial patterns of perception (Baum 2021; Hemmerling et al. 2020). Since 2010, the relocation of the island's inhabitants to the vicinity of the city of Houma has been taking place, for which 48.3 million dollars of state funds have been made available (Jessee 2020). The resettlement of the residents is largely taking place in a self-contained structured manner in order to avoid social disintegration through the loss of locally formed social capital. In addition, the people living on Isle de Jean Charles were intensively involved in the planning of the project (Jessee 2020). Here it becomes clear—in the landscape theory interpretation used here—of what Jesse (2020) noted with regard to dealing with the social consequences of land loss, that not only is Landscape 1 historically conditioned, interpreted in the Mode C, and the subject of projections (Landscape 3c), but that local societies struggling with the loss of their Landscape 1a, which they themselves are not responsible for, have a right to continue living together, if this is desired.

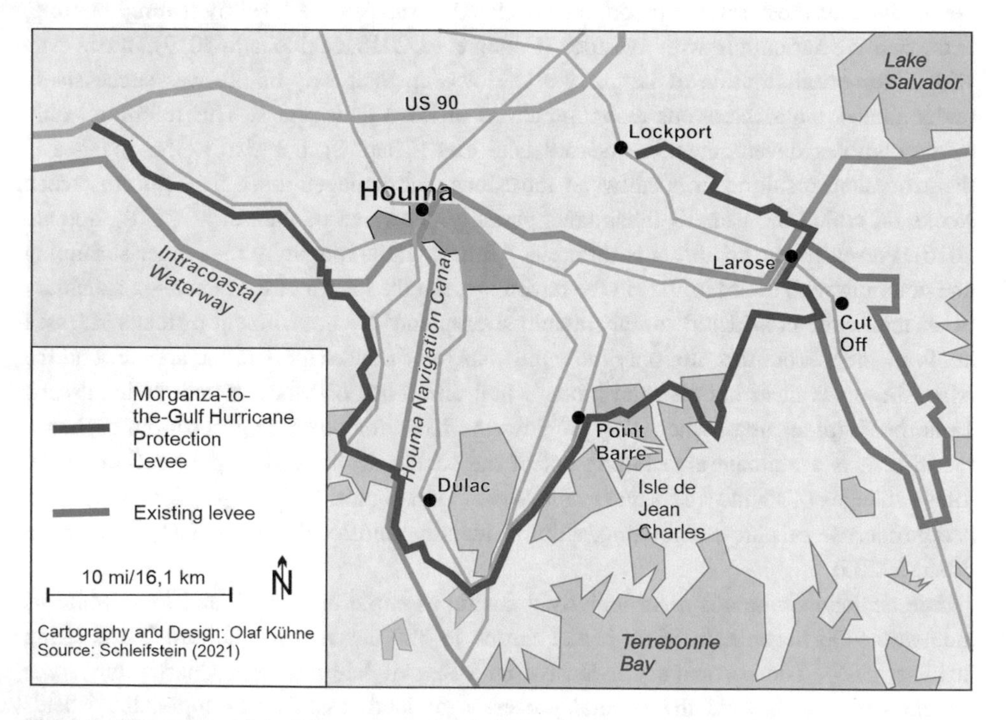

Fig. 6.1 The course of the Morganza-to-the-Gulf Hurricane Protection Levee System. (Own illustration, on the basis of: Schleifstein 2021)

Media coverage of Isle de Jean Charles and its (former) inhabitants focuses simultaneously on two concrete events, primarily upon the impact of hurricanes and concurrently on the issue of climate refugees (Baum 2021; Herrmann 2017) which is linked to a victimization of the population that in turn entailed the implementation of disempowerment and is linked to a distancing of media users from the local population. Initially, the events were contextualized less intensively in relation to anthropogenic climate change, albeit ultimately, the imagery used conveyed a picturesque antiquity aesthetic, characterized by dilapidated infrastructure and abandonment (Herrmann 2017) as it has been cultivated in landscape aesthetics since the eighteenth century (Hauser 2004). The study by Linda Baum (2021) in which she analyzed the discourse of 28 newspaper articles on Isle de Jean Charles and its (former) inhabitants, shows a clearer contextualization of the relocation measures as a consequence of anthropogenic climate change, whereby this was discussed much more in national newspapers than in the regional newspapers, whose reporting was more influenced by everyday life. Also, the coverage of the national newspapers was more connected to the ecosystemic consequences, the associated loss of livelihoods of the population in terms of agriculture, fishing, and hunting, as well as a diminishing social and cultural life on the island. The inhabitants of the island would have their say, especially "when they complain about something, such as the critical situation on the island or the dissatisfaction with the course of the resettlement process so far, from which they have long felt excluded" (Baum 2021, p. 163) which in turn documents the diminishment of the residents to a victim status.

If anthropogenic climate change is increasingly addressed in newspapers as a major cause of land loss in Louisiana, with its sometimes-fatal consequences for local communities, the categorization by discussants in social media is much less clear. On the contrary, the discourse positions on the 'existence' of anthropogenic climate change are almost equally divided between those adhering to the position on climate change almost invariably shared by scientists and those rejecting it (Kittler 2021). Whereby the dispute between the antagonistic discourse positions is conducted with great vehemence, often bypassing the rules of respectful communication, but with recourse to the moral subordination of the other party to the conflict—by both sides, including those claiming to be fact-based (Bishop 2014; Brunnengräber 2018; Kittler 2021; Kovaka 2021; Kühne et al. 2019; Neverla et al. 2019).

In order to expand the state of knowledge concerning the social construction of the processes surrounding land loss in Louisiana in mass media discourse, in this case on social media, we will devote an empirical investigation to the question of how the consequences of Hurricane Ida (August 2021) are discussed. For this purpose, we resort to the analysis of an Internet video together with its comments, a procedure that is becoming increasingly widespread with regard to the observation of the social construction of the world in general and of landscape in particular (Kavitha et al. 2020; Kittler 2021; Kühne 2012a; Kühne, Koegst et al. 2021; Kühne and Schönwald 2015; Kühne and Weber 2015; Madden et al. 2013; Mann and Rauscher 2021).

6.3 Social Media Communication of Threats to the Louisiana Coast—YouTube Video: "Grand Isle, La Drone Video of Hurricane Ida Damage Whole Island- Category 4 4k" Including Comments

Acknowledging the dominant importance of YouTube, a platform founded in 2005 and acquired by Google in 2006 (Geimer 2019), we use a video published on this platform for our analysis of the topic involving the media construction of the threats to the Louisiana coast. After all, YouTube has managed to make its way from a niche program for young people "into the center of globalized media culture […] at breakneck speed. Its offerings now range not only from the infamous cat and baby videos to lectures on the origins of the universe or interviews with contemporary philosophers, but almost all areas of everyday culture as well as specialized professional worlds are reflected" (Geimer 2019, p. 2). With the penetration of everyday life by video, YouTube can be seen as a major driver of the mediatization of large parts of society, which certainly qualifies it as a central object of scientific investigation of current social processes (Burgess and Green 2018; Geimer 2019; Kühne 2020b; Tuma 2018).

YouTube not only offers the opportunity for scientific analysis of the videos available on the platform, but also the comments and discussions published under these videos can provide interesting insights into media communication. For example, in this analysis we looked at the comments posted under the video "Grand Isle, La Drone video of Hurricane Ida Damage whole Island- Category 4 4k" (WXChasing 2021). The video shows various drone footages of Grand Isle taken in the immediate aftermath of Hurricane Ida's passage in the summer of 2021. Different sequences of the flyovers are cut one after the other without commentary or background music. The damage and destruction of buildings and infrastructure on the island visible in the footage triggered an intense discussion in the commentaries.

A total of 488 comments were analyzed qualitatively to examine the discussion about the effects of the hurricane on Grand Isle. Conforming with the content analysis according to Mayring (2008) the comments were first assigned to an inductively created code system with regard to recognizable classifications and attributions, from which patterns of corresponding classifications could then be derived. Only one code was assigned to each comment; if a comment dealt with several aspects, the more dominant one was considered. In this way, a total of 13 interpretation patterns could be identified. In the following paragraphs, the individual interpretation patterns are presented correspondingly and analyzed with regard to the theoretical foundations. No further attention is paid to the pattern of 'references to the video' (a total of 45 comments), because comments that are summarized here only marginally have to do with the effects of Hurricane Ida on the Grand Isle, rather it is particularly about reactions to drone footage, the pilot of the drone, the drone used, and its control. Often, the end of the comment is followed by the addition of condolences such as "praying for all" or "so sad". Similarly, in 64 comments, no further meaning to video intake or content could be identified, so these comments are also disregarded in the following presentation. From the remaining comments evaluated, we identified eleven patterns of interpretation. These are examined in more detail in the following paragraphs and clarified with the help of example comments, which are shown in text boxes. We grouped the interpretive patterns into three groups according to their content[1]:

'Expressions of sympathy and encouragement': A total of 25 comments dealt predominantly with expressions of sympathy and encouragement towards those affected on Grand Isle (see, for example, K30; K226).

'People on Grand Isle': 29 of the comments analyzed dealt with people on Grand Isle, with commenters inquiring whether people stayed on the island during the hurricane, how people were doing on the island, or whether there were any fatalities (see K420, among others).

'Reports and Experiences': A total of 44 commentators reported on their own experiences and impressions of hurricanes in general and Hurricane Ida in particular; the comments were apparently written both by people with a closer connection to Grand Isle—including residents (such as K9)—and by people from all over the world who have had their own experiences with hurricanes, with comparisons also being drawn among other hurricanes in the past or in other places (e.g. K228).

'Costs/Insurance': 15 comments dealt with the costs incurred by the destruction and damage to buildings and infrastructure caused by Hurricane Ida. Insurance was mentioned particularly frequently, as its use in affected areas would increase premiums for everyone (see, for example, K176).

'*Climate* change': Climate change is only mentioned in four comments, in which it is held responsible for increased costs and seen as a cause for increased and more intensive occurrence of hurricanes, whereby a dystopian version of the future is drawn, especially for the coastal regions, and only limited life opportunities are attributed to residents (see K104, among others).

'Reactions to the video': With a total of 34 comments, various reactions are summarized in this pattern that relate directly to the video, for example, reference is made to the amount of sand that extensively covers the surfaces, and impressions of commentators are put into words (e.g., K285). There is also a certain disbelief that people will voluntarily live on the island and rebuild their houses, but at the same time there is optimism and confidence that the inhabitants of the Grand Isle will certainly succeed in making it habitable again.

'Comparison with war scenes': In seven of the analyzed comments, comparisons of the scenes depicted in the video with theaters of war could be identified (see, for example, K175); in particular, the rubble lying around and the partially heavily destroyed houses seem to evoke associations with the explosion of a bomb (e.g., K242).

[1] We have reproduced the comments here verbatim without correcting the spelling, as this reflects the usual nature of communication in social media.

Text Box 1:

Comments of the first samples

"*Breaks my heart to see Grand Isle this way* ⍰" (K30);

"*Hang on everyone*" (K226);

"*I heard everyone was alive, though. Things can be replaced, people can't.* ⍰⍰" (K187);

"*Are the people that stayed behind on Grand Isle ok??? [...]*" (K420);

"*3:09 that was my camp the big hole in the ground and 1 palm tree. It was my step-dads camp before me, it was built in the 60's sadly we left all the photo albums in it when ida struck. its all gone. I cant believe it*" (K9);

"*Looks like Mexico Beach after Hurricane Michael, just slabs of concrete what were houses*" (K228);

"*Insurance will just pay for it which will raise everybody's rates they'll just build right back there again and it'll just happen again and again with no lesson learned*" (K176);

"*[...] considering the weather is going to get worse because of climate change... and we are NOT going to be able to stop it with what the world is currently doing, no matter what America does... these people should not be allowed to rebuild since we know this will happen again, and again... how many times, already, have they been flooded out or blown away? [...]*" (K104);

"*Hard to wrap one's mind around so much devastation. Harder to believe people will choose to rebuild in this area*" (K285);

"*This looks worse than a war zone, pray for these people, to be able to recover. Such destruction*" (K175);

"*Looks like a bomb went off.....*" (K242).

In addition, patterns were identified that relate to, discuss, and at times even vehemently question the reconstruction of Grand Isle (see Text Box 2 for more on this):

'Building advice and ideas': A total of 82 comments deal with structural aspects and suggestions. In particular, a poor structural design of the buildings is criticized (43 comments), which according to the commentator is because of the use of wood (see among others K51). To remedy this, better engineering performance is usually suggested or leaving the island is mentioned as a last resort (among others, 52). The use of more stable building materials is also echoed by commentators in terms of speculation about why some houses are more damaged than others, while still others appear to have little damage (See K239, among others; 25 comments total). In addition, other commentators (14 comments) argue that stricter building codes on the island would have ensured less damage to newer homes in particular (see K92, among others). At the same time, however, there is also discussion about whether these are strict enough (K154, among others).

'Do not rebuild': a total of 62 of the comments wonder why the island is built on at all (37 comments; including K78) or oppose rebuilding the buildings on Grand Isle for various reasons (25 comments). Others believe that reconstruction should only occur if certain standards are met that could protect the buildings in the event of renewed flooding (including K225). In addition, some commentators do not understand why the island is built on at all, given that Grand Isle is located in a hurricane zone directly on the coast and is therefore exposed to natural hazards (K370). Likewise, some commentators imply that the inhabitants are to blame for the fact that their houses were damaged or destroyed—they should not have built so close to the coast (e.g., K158).

'Vacation camps only': 29 of the comments analyzed discuss the claims of some commentators who see the damage to Grand Isle buildings by Hurricane Ida as not regrettable because the island is mainly home to vacation camps owned by people with high endowments of symbolic capital (including K357). Emotional ties and the presence of permanent residents on the island are completely disregarded. These assertions are contradicted by commentators who describe that people who work as "essential workers" with an income "below or at the poverty line" live on the island (K287).

Text Box 2:
Comments concerning the reconstruction of the island

"If America didnt have wooden houses and stuff, it might have withstood it Ok. Cant believe houses are so expensive there but made out of wood. You cant expect anything but destruction in hurricanes." (K51);

"Build back better using steel, concrete or leave" (K52);

"I'm more surprised at the number of houses that appear more or less untouched than the number that are completely destroyed. With the right materials and methods you can build a house that will take a very serious beating" (K230);

"Ever wonder why one building has been COMPLETELY destroyed while a building 100yrds away has MINOR damage? Building codes" (K92);

"Louisiana lacks strong building codes, but some folks saw fit to build to higher standards" (K154);

"Doesn't take a rocket scientist to determine you shouldn't build houses on a barrier island in the Gulf of Mexico.....but people will...Merica!" (K78);

"There should be compensation for losses, but they should not allow this island to be rebuilt unless the locals build to standards that would involve lots of concrete and rebar and pay for the cost of rebuilding the infrastructure through their property taxes. It would be nuts to do otherwise" K225);

"You think people know that [...] live in a hurricane or tornado region would build elsewhere..." (K370);

"When you build your home or business this close to the water you are asking for trouble. Everyone has flood insurance which means the government will pay to

rebuild their homes. My advice [...] is sell the land and move inland. It is not worth it" (K158);

"Don't feel sorry this is rich peoples toys" (K357);

"These comments are unbelievable. Blaming "rich people" for the houses. Most of the people who live there are at or below the poverty line. These houses are meant for those essential workers" (K287).

In addition, three more patterns could be identified, whose basic understanding of the world is beyond generally accepted scientific knowledge (see Text Box 3):

'Conspiracy theories and Mother Nature': In only four of the comments examined, connections to conspiracy theories could be identified, with the modification of the weather by humans in particular being held responsible for the hurricane (e.g., K128). Likewise, other commentators (7 comments) personify nature with terms such as 'Mother Nature', which is beyond scientific considerations (e.g., K353).

'Religious references': Smaller religious references such as in K338 or similar could often be found towards the end of comments, but religious references were predominant in 37 comments. In some instances, these are references to Bible passages or other mentions where no direct reference to the Grand Isle can be made, as well as prayers for those affected (such as K199). However, there are also clear statements about the hurricane's impact on the island being God's will (K89; K137). In the responding sub-comments, discussions arise that start with statements such as the one just mentioned, but then take a completely different direction evolving into fundamental discussions about the existence of God and what he might look like which are no longer relevant for this evaluation.

Text Box 3:

Comments whose basic understanding exists outside of generally accepted scientific bodies of knowledge

"Weather modification did this." (K128);

"A few more like this, LA will be majorly redesigned by Mother Nature..if not removed from map" (K353);

"God bless you all" (K338);

"Praying for the people of Louisiana" (K199);

"Watch every one rebuild on the same spot, with very very expensive, nuclear hardened, new homes, only for God to tear them down again. They just don't get or comprehend the picture don't they. He has adequately warned us to change our wicked behavior" (K89);

"nice thought but I think the lord has forsaken this land" (K137).

Overall, only four references are made directly to climate change and its consequences in the analyzed commentaries; 'alternative interpretations' beyond the scientifically established findings are somewhat more strongly represented, but these also play a rather subordinate role. The focus of the discussion is rather on everyday experiences and pragmatic reactions to the destroyed buildings of Grand Isle: 'How can we build better?', 'Should the island be settled at all?'. Various suggestions, possibilities for improvement and ideas are put forward, but at the same time also accusations and insinuations, even provocations. Reasons for destroyed or damaged buildings are predominantly sought in structural and constructional aspects, only some refer to the aspect of climate change. No mention in this context is made of the oil industry and its involvement in wetland destruction, land loss, and the related greater destructive potential of hurricanes when they hit coastal areas. Other causes of land loss in Louisiana, outlined in Chap. 3, do not resonate in the comments. In the majority of the comments, only limited abstractions can be identified, such as climate change or even alternative interpretations up to conspiracy theories; rather, people react to visual impressions of the video and refer to them.

Especially with regard to the form of reconstruction, some commentators, who according to their own information are active in the construction industry or crafts, have their say, but also there are some who offer general advice and describe observations from the video or mention adaptations and structural adaptations from other areas impacted by hurricanes. There is little reflection on the fact that such a discussion in this setting—a comment section on YouTube—is hardly relevant to the reconstruction of the island; reference is made only to the establishment of certain building codes from an administrative aspect. Accordingly, it becomes apparent that some of the commentators are of the opinion that they possess 'expert special knowledge' regarding hurricane-proof construction, but whether this is actually the case cannot be determined on the basis of the analyzed comments, therefore it is possible to speak of a supposed Mode C in this case.

Also, clearly evident is the personal connection to Grand Isle, with some of the comments written by people personally affected or involved. In addition, some speak up who have had similar experiences in other areas. Here, the extension of the concept of symbolic capital according to Bourdieu (1984, 1985, 1989) can be seen (as we have discussed in Sect. 5.2), as illustrated with particular clarity in areas affected by hurricanes, regional subsets of symbolic capital are formed that are adapted to the spatially oriented climatic conditions. These are particularly evident in the analyzed comments involving the diverse proposals and experiences for constructing and securing hurricane-resistant buildings. Accordingly, Mode B is also regionally bound and not socially universal. This is particularly evident in the United States, which is characterized by subsocieties (cf. among many: Currid-Halkett 2021; Dahrendorf 1963; Soja 1993). Within a spatial and climatical context, the different interpretations become apparent, for example, in the juxtaposition of regions of the United States that are heavily impacted either by water (rain, flood, sea) or snow (blizzards, lake effect, etc.): In response to the comment "I'll stick with Lake Michigan in the summer and howling winds with 2–3 ft of snow in the winter!" (K351) is met with slight irony: "Most of us would rather have 10 hurricanes than six inches of snow lol" (K352).

In contrast, some of the comments blame the residents of Grand Isle themselves for the destruction; after all, the place of residence could have been chosen outside the hurricane zone and not in the immediate vicinity of the sea. This also shows the reluctance of many commentators to use government funds to rebuild coastal regions: "This place should no longer be inhabited, the storms come harder ever year. The same money it would take to clean up and be rebuilt, should be used to relocate citizens to a safer place. God Bless you all" (K338). Accordingly, for some, there is such a strong monetization of the reconstruction of hurricane damage that there seems to be no willingness to indirectly support the rebuilding of destroyed homes—and thus the homes of those affected.

Simultaneously, a phenomenon becomes apparent: In some sections, commentators are aware that as a barrier island Grand Isle is of importance for coastal protection, especially during hurricanes which elevates its particular importance for coastal protection. Such comments present some scientific understanding of the region, thereby suggesting the presence of an interpretation of what is perceived in Mode C. Equally, however, the administrative side of the state is accused of a failure in that the island has been built on at all. Here, ignorance of the fact becomes evident that Grand Isle has not just recently been inhabited, but has been by Native Americans for several generations (Town of Grand Isle 2018).

One comparison that stands out among the analyzed commentaries is with the tale of the three little pigs—an English-language nursery rhyme in which a wolf blows down the two unstably built houses of straw and wood, respectively, of the first two little pigs, but fails with the third because the latter built his house out of bricks (cf. i. a. Flora Annie Steel 1922). The comparison is very direct: "Wood and nails—vs Bricks and rebar—Three little pigs' kind of told us about it in their story." (K56). In other words, if the buildings on Grand Isle had been more stable, the hurricane would not have destroyed them. Accordingly, the inhabitants of the island are also clearly assigned a share of the blame for the destruction of the buildings and in some comments even denied the ability to adapt and learn: "And the pigs will just keep rebuilding and let the big bad wolf blow their house down again. Rinse and repeat." (K33). Provocations like these, however, also show the limits of the approach we have chosen, because in these neither 'personal experiences' (Mode A), 'social knowledge' (Mode B), nor 'expert special knowledge' (Mode C) can be recognized, rather it is judged from an high-minded position, provoked, and repeatedly made clear to the inhabitants of Grand Isle that life chances on the island are—if at all—only available in a small form. Repeated statements that Grand Isle is not a place for permanent habitation, such as: "It has been told and told again & again. Do not build your home on sand" (K191), to direct insults and condemnation of residents (including "These are the kind of people who need a blowtorch to the face 100 times to learn fire burns sorry zero sympathy. LEARN!!!!!" K123), being in the end result 'thou shalt' commands of externally directed moral ligatures. The exaggeration of one's own worldview recognizable in these comments and the moralization associated with it leads to a discrediting of the residents of Grand Isle, in this example by means of an insinuation that the residents are inferior and unable to learn from the past (see in more detail Kühne, Berr, and Jenal 2022). However, not all commentators see the destruction of the buildings as a sign of a non-existent learning ability of those living there; rather, the patriotism of Americans to cope with everything is revealed: "[...] I'm praying for all evolved, but sometimes mother nature, being devastating, brings out the American in us.... they will rebuild, they will prosper, and anything anyone can do to help out is the biggest blessing rite now" (K391). Or, put more succinctly, "Built Back Better" (K337). This is also evident in comments apparently written by people who are native to Grand Isle, revealing a deep connection to their home: "We're in progress of rebuilding. This is home" (K343).

Some commenters think the damage on Grand Isle is not that bad because "Most of these are camps not houses" (K339). The term 'camp'—a local term for small houses, conveys what are "often more like lavish fishing vacation homes with docks and modern amenities" rather than, say, cabins or tents (cf. Jessee 2020, p. 160)—is used in this context by some commenters as a pejorative term for the vacation homes of people with high endowments of economic capital, up to and including cabins or camping buildings. A fact that also leads to discussion among commentators ("'camps'? What are you talking about? Do you know what the word means?" K386) and several attempts are made to correct this assumption: "You have posted this nonsense so many times that it is getting ridiculous. Yes, there are camps on the island but there are also lots of permanent homes. Have you ever spent any time there? Recently?" (K408). The comments of one person in particular stand out, which were posted from the perspective of Mode A in similar wording a total of six times in response to comments that can certainly be understood as provocative: "Grandisle is not a glamorous "beach ocean view" area. This area is mainly for essential workers like those who work on oil barges, fishermen, scientists, marine biologists, fish and wildlife, and other agencies. More than half of the people who live here are at or below the poverty line. Last I checked the average cost was around $130,000 for a house there. An ocean view in Florida is millions. People are so ignorant to the culture of this area. Wow!!!" (i. a. K360; in this case, a response to the comment, "Don't feel sorry this is rich peoples toys" K357). This assumption that the destruction of the homes of wealthier people is less serious than when homes of less wealthy segments of the population are damaged or destroyed illustrates a 'vulgar class struggle' that both draws attention to the division in society and also concurrently suggests that despite the recurring destruction of buildings on Grand Isle, there is envy toward those who live there or own vacation homes.

6.4 Referencing Back to the Theoretical Framework

The communication presented in the previous section exploring the comments on the YouTube video "Grand Isle, La. Drone video of Hurricane Ida Damage whole Island-Category 4 4k" provides a valuable basis for interpretation using the theories and theory extensions drawn upon, including in light of media content analyses with comparable thematic focuses (such as: Kühne, Koegst et al. 2021; Kühne and Weber 2019). The relevance of Ralf Dahrendorf's extension of the concept of ligatures to include corporeal ligatures and corporeally mediated social ligatures becomes clear: the communication repeatedly comes back to the vulnerability of human beings as a result of their physical constitution. The discussion about designing buildings with a higher resilience to hurricane and flood events can be interpreted as an attempt to minimize corporeal ligatures emanating from World 1. In turn, the discussion about state coastal protection measures in Space 1 or the sense of inhabiting the barrier islands can be understood as a struggle for corporeally mediated social ligatures, because—any change in coastal protection measures or the state-ordered abandonment of residences—has visceral and physical consequences for the people residing there, consequences that emanate from social (more precisely political and planning) decisions. Here, then, the logic of the spatial or landscape Mode C dominates that of Mode A and B. The distinction of the (in this case, possibly self-proclaimed) Mode C reference to Grand Isle's Space 1 from the locally and regionally specific social and incorporated cultural capital of both Mode A and B is also evident in some comments when references are made by people outside of coastal Louisiana for a measure of appropriate management of local and regional hazards.

Compared to our investigation of communication by examining a YouTube video on the flood disaster in the Ahr Valley (Germany) in July 2021 (Kühne, Koegst et al. 2021) it is possible to identify (exploratively) some specifics of the communication regarding the video studied here. The discussion about the Louisiana video is strongly focused on the concrete, often related to typical individual or (this can be concluded from the contextualization) professional experiences. This is clearly different from the Ahr Valley video, where abstract discussions about climate change and the political questioning of flood protection, which often take on a life of their own, are clearly more relevant. Connected to this is also a different direction of the discussion: In the comments on the Louisiana video, the question of how the individual person, or household, can mitigate its corporeal ligatures is discussed more frequently, through individual measures, while the comments on the Ahr Valley video are more strongly aligned with the question of culpability for the events (prominently: emissions of greenhouse gases). Thus, there is more discussion regarding Grand Isle on the issues of increasing or maintaining options (while repudiating socially mediated ligatures, both physical and corporeal), whereas in the Ahr Valley video there is a strong focus on communicating externally directed moral ligatures. Here, World 2 is not conceptualized as a starting level for an innovative and self-responsible dealing with World 1 and 3 (as dominates in the Louisiana video), but as an object of a pathological attribution of guilt (cf: Ackermann 2020; Grau 2017, 2019; Kostner 2019). In large portions of the comments on the Louisiana video, there is a stronger focus on the factual level, combined with a (critical and conflictual) discussion of the arguments of the other side. For the most part, the discussion is pragmatic, based on the search for the more suitable argument. In contrast, the discussions about the Ahr Valley video quickly develop from a conflict of facts to conflicts of identity and values, which, attributable to the high degree of moralization, can hardly be regulated, instead contributing to the further radicalization of communication and recursive solidification of milieu-specific patterns of interpretation and values (for more details, see: Berr 2017; Kühne 2018b; Luhmann 1993, 2016; Sofsky 2013; Stegemann 2018). Comparatively, the comments on the Ahr Valley video focus much less on the availability of economic capital than in the Louisiana video (here, especially in the form of references to the impacted vacation residences of wealthy segments of the population). The reasons for these different patterns of interpretation and communication may lie in a stronger prevalence of pragmatic views in the United States compared to a wider prevalence of ethic attitudes in Germany (on this: Holzner 1994, 1996; Schneider-Sliwa 2005), which is also suggested by other empirical studies (e.g.: Kühne 2012a; Kühne and Schönwald 2015; Kühne and Weber 2019; Wagner 2019), although a systematic cross-cultural comparative study of this is still pending.

7 Terrestrial Marine Hybrids on the Louisiana Coast—A Phenomenological Access

Having explored Mode A and Mode B aspects of coastal land loss, using the example of Grand Isle, we will now turn to our own experience of this 'extreme' Landscape 1. In doing so, we will refer back to the 'new phenomenology' of Schmitz, who accordingly emphasizes the impacts of phenomena upon who physically exists and experiences the world (Schmitz 1980).

7.1 The Phenomenological Walk—Some Basic Considerations

The focus of our study is the Grand Isle region of Louisiana in May 2022, a place that defies classic tourist stereotyping in its current state of Landscape 1 as a result of recent persistent destruction due to hurricane Ida in 2021. Notwithstanding, in the spirit of representing a synaesthetic experience, we wanted to include non-visual perceptions along with our in person account in order to represent it more vividly and connect it to non-reductionist notions of landscape (in more detail: Bahr 2014; Bischoff 2005, 2007; Edler, Kühne et al. 2019; Edler and Kühne 2019; Endreß 2021; Kühne 2018c; Kühne and Edler 2018; Raab 2001). The experience of Space 1 as Landscape 1 takes place with us exploring Grand Isle as an 'other space' (Foucault 1990), thus a space 1 that challenges stereotypical patterns of interpretation and evaluation (Mode B), that has hardly any parallels with our typical German Mode A (central Ruhr area and southern environs of Stuttgart), and a space 1 that was additionally difficult to anticipate in its severity even after extensive Mode C engagement involving the processes underlying land loss. Such an arrangement of Landscape 1 makes individual references exploratorily meaningful and requisite to fathom the ambiance (cf. Hasse 2007, 2012; Kühne, Jenal, and Berger 2022). Such considerations lead us to Herman Schmitz's 'Neue Phänomologie'—New Phenomenology (Großheim and Kluck 2010; Schmitz 1980). This philosophical approach binds the individual experience to its physical sensation as the primacy of

O. Kühne and L. Koegst, *Land Loss in Louisiana*, RaumFragen: Stadt – Region – Landschaft, https://doi.org/10.1007/978-3-658-39889-7_7

interest, whereby the unfamiliar gives challenge to the experience in a special way. Experiencing situations can be understood as a complementary approach to the (positivistic) comprehension of spatial constellations (Großheim 2010). Such an approach is bound to the physical sensing of situations. Therefore, it requires the physical presence of the sentient human being within the Landscape 1 that is being engaged (Ljunge 2013). Since such an experience of landscape reaches beyond what a remote sensing description of Space 1 allows, it requires changes of location specifically without using such high-tech options (Foxley 2010; Macpherson 2016). Devices and methods which ultimately cause a distancing from Landscape 1 by hastening and exempting us from the first-hand efforts of developing spatial understandings (Kühne, Jenal, and Berger 2022). Correspondingly, we use the method of the 'phenomenological walk', in which impressions are noted describing how Landscape 1 is experienced with all the senses (Burckhardt 2006; Kühne and Jenal 2020a; Wylie 2005). The method of the phenomenological walk can be distinguished between the methods of the 'dérive' (for instance at Diaconu 2010; Smith 2010) on the one hand and that of 'Science of Walking'/'Strollology'/'Promenadology' (Burckhardt 2006b) on the other. Compared to the 'dérive' method, the phenomenological walk is characterized by a more mindful purposefulness. The element of 'letting oneself drift' takes a back seat to a stronger elaboration of the route planning based on prior technical knowledge as well as the engagement with maps and aerial photographs. Compared to the science of walking, phenomenological walks differ in their normative neutrality. Thus, they do not claim to impart new spatial perspectives to third parties present (e.g., by means of interventions). Also, they are not—like the science of walks—intended as a method of preparing or planning a route.

The starting point of our phenomenological walk is, reminiscent of the classic texts dealing with vernacular landscapes by J. B. Jackson (Jackson 1984; Lefkowitz Horowitz 2015), a central place of the area—in the absence of a bus or train station, the main street. As a locale, we witness a materially and spatially narrower Landscape 1, as well as in terms of aesthetic, emotional, and normative access. This engenders a special manner of individual experience (for the discussion of space and place see among many: Furia 2021; Jiraprasertkun 2015; Massey 2013; Scott and Sohn 2018; Tuan 1989).

The separation of the phenomenological impressions we collected on Grand Isle into urban and beach areas is not clear at the scale of an island like this one, since urban or residential community and beach are spatially very close to each other and without clear boundaries, for example resulting from the omnipresence of sand—which is stereotypically associated with beach rather than residential properties—seem unclear and cannot be clearly separated, especially with regard to multisensory perception. The only form of obvious, visually recognizable demarcation between beach and community is the levee which is supposed to protect the residential areas from waves and water coming from the Gulf of Mexico, but even this is no longer clearly recognizable in parts and in some places is even completely destroyed. Nevertheless, we distinguish in the following representations of our impressions of both Grand Isle settlement and beach, in order to clarify the difference between the inhabited area and the beach area, in the preformed stereotyped expectation. Furthermore, as will become clear in the following explanations, in the community areas we are constantly confronted with the visual structural features as well as the effects and damage of Hurricane Ida, while on the beach, with the view of the sea, the damage recedes a little into the background and we get an impression of an island at the edge of the Gulf of Mexico that under normal circumstances seems almost paradisiacal, even if the oil production platforms on the horizon cause the idyllic picture to unravel a little.

7.2 The Settlement

We begin our phenomenological walk at our lodging on Grand Isle along the main road that runs the length of the island. After driving onto the island alongside Bayou Lafourche—on whose shores we have already been able to get a clear impression of the hurricane's impact—and, the further south we get, along the elevated roads, we are further impressed by the lengthy drive across the coastal marshes, which, the closer we get to the island, are more and more frequently fully covered by water, making us all too aware of the coastal land loss that was familiar to us from our Mode C literature study (see Fig. 7.1). Shortly before reaching the island, we can already guess what awaits us: On the horizon we can see a destroyed large hall in the harbor area of Port Fourchon, directly next to it a new building is being erected. It was not apparent to us why structures of this size are established at a place which is hit so frequently by hurricanes. When we finally reached Grande Isle, we got our first view of the island and its elevated buildings from the elevated road that leads from Elmers Island to this island. At the roadside, we see a sign with the words "Jesus Christ reigns over Grand Isle" (see Fig. 7.1). We are a little surprised and uncertain interpreting the statement of this sign regarding whether Jesus, or God, is stereotypically seen as a protector or as a tyrant. Especially regarding the idea that the destruction on the island was caused by Hurricane Ida (often considered an 'Act of God' in US culture and insurance industry; see among others Steinberg 2006), a certain ambiguity arises for us here. The statement of the sign intensifies our tension as to what will await us once we have left the car and look around the island on foot.

In the following paragraphs we will present our experiences according to the sensory impressions and summarize them correlating the atmosphere felt by us. The description of our impressions will be supplemented by photos, which should serve particularly supportive of the descriptions. The description of visual impressions will be specifically supplemented by the presentation of acoustic and olfactory stimuli.

The elevated houses, which are located on short side streets branching off at right angles from the main street, are particularly striking. The plots of land are irregularly built on. It remains difficult to comprehend that there once stood houses on these empty lots, some of which have been neatly cleared. The remaining buildings show varying

Fig. 7.1 The elevated road to Grand Isle crossing both marsh and water impresses us on the way to the island (top left), there we see elevated houses in various states of damage (bottom right). In parts, entire house sections are missing and we can see which rooms were used as bedrooms or kitchens (bottom left). The sign at the beginning of the Grand Isle astonishes us a little and makes us think (top right). (Photos: Lara Koegst 2022)

degrees of damage, with the degree of damage increasing towards the south and west from houses with minor damage (for example, missing parts of the facade or destroyed window panes) to completely gutted buildings, where we have a glimpse of the entire interior—or what remains of it eight months after the hurricane—preserved or only the basic structure recognizable (Figs. 7.1 und 7.2). The sight of these destroyed buildings, where a normal life is so easy to imagine based on the direct view of furnishings and other possessions, is depressing and illustrates the power of the storm winds and the massive amounts of water. The impressions remain inconsistent, at times contradictory: The idea of how things might have turned out on the island during the hurricane is apparent in view of the destroyed houses. The force of the storm and the massive quantity of water is particularly evident in the height of the elevation of the buildings. The many posts, connecting the living area of the houses with the cement foundation, look like a bizarre forest (Fig. 7.1). In some areas, the remains have already been cleared away, so that only empty elevated platforms signify the original buildings. In between, especially along the main street, rubble and debris have been cleared and gathered into large piles—reconfigurations of the houses are created in which household items and building components can be recognized (see

Fig. 7.2 In the residential areas of Grand Isle some houses led us to assume that hardly anything has changed since the hurricane in 2021, since the remains of the houses were not cleared away or demolished (above right), at other houses it is clearly recognizable due to large debris piles at the side of the street that some has already been cleared up and repaired (above right). Particularly evident are the height differences with which new houses are built, some of them being high enough to allow trailers and mobile homes to be parked underneath (below). (Photos top right Olaf Kühne 2022; top left; bottom Lara Koegst 2022)

Fig. 7.2). In some of the houses we have the impression that they are still in the same condition in which the hurricane left them, in others it is clear that some of them have already been rehabilitated, renovated, and even completely rebuilt. We are confronted with an alternating mosaic of buildings and building remnants, houses in different degrees of devastation, new and renovated houses as well as foundations from which the buildings have been cleared. The newly constructed houses are thereby built on higher platforms (see Fig. 7.2)—about two stories high—evidence of the stricter building regulations. The new elevated buildings are so high that the trailers and mobile homes that we repeatedly see on the island can be parked underneath. The closer we get to the school and the multiplex center of the island; the more vegetation is visible. While we have encountered mostly ruderal vegetation so far, palms and other trees grow here in some locations, and we can guess from the rather systematic layout of other plants where gardens may have once been laid out. Signage that we see on some of the cleared plots of land seem remarkable to us: reference is made to the prohibition of removing sand. How important sand is for the island becomes clear to us here once more. Only a few cars drive past us, and we also encounter few people. We have to smile when seeing those who are driving around the island in golf carts with off-road tires, as we tend to associate these vehicles with the lush green, well-kept lawns on golf courses rather than the sandy areas of the Grand Isle.

The salt in the air—coming from the nearby Gulf of Mexico—is also gustatory present in the settlement. A taste that reminds us of beach vacations and is therefore in great contrast to the so visually present outcomes of the hurricane. Tactilely, we can feel the wind in the air and the sand under our shoes, which seems omnipresent—even on the paved roads. Constant companions of our walk are the humid air on our skin and the hot temperatures, where even the rain, which fell briefly at the beginning of our stay on the Grand Isle, hardly cooled us down.

Olfactorily, the sea and the salt water are also very present and clarify the immediate proximity of the Gulf of Mexico. Furthermore, we repeatedly smell the exhaust fumes of cars that are on the road without catalytic converters or whose engines burn considerable amounts of oil in addition to gasoline. From time to time, we can smell decaying fish, especially when we get closer to the beach, often accompanied by the smell of garbage, which decays quickly in the humid air and the hot temperatures. Equally present are the olfactory stimuli of putrid water that collects in puddles along the roadside. In addition, near the accumulations of houses and debris, rotting wood makes itself pungently impactful. In contrast, in the vicinity of renovated houses or those under construction, we smell new, freshly sawn wood, and sometimes drying paint. The two contrasting olfactory spheres once again underline our already visually appraised impression of the different ways of dealing with the destruction of the houses: Between newly built and the still existing unchanged damage caused by the hurricane, all stages can be a multisensory experience for us. We are impressed by the will of the people to repair their houses again and again, to rebuild them, and as we could guess from the glimpses we got of the inventory, to reassemble household goods and similarly to lose these documents of memory, like pictures or photo albums, again and again.

There is widespread silence, especially barely perceptible sounds of human provenance. What is acoustically present, however, is the rustling of the wind, our own footsteps on the streets lightly covered with sand, the crying of seagulls, and, near the sea, the crashing of waves onto the beach. All of these sounds correspond to the stereotypically positive acoustic stimuli we expect to hear near the beach, and in turn evoke memories of beach vacations for us. However, we also perceive other sounds at certain points, for example, we hear the engines and tire noise of the cars and golf carts that drive past us, and in one of the elevated, damaged houses we hear the loud sound of music from the current charts. This, in turn, has a somewhat irritating effect on us, as it stands in clear contrast to the damaged houses. But it also makes clear that life on the island goes on even after Hurricane Ida and can now be celebrated again—a struggle for normality. The audible hammering of repair work on individual houses reinforces our ambivalent impression of not wanting to be defeated, of the confidence of the island's inhabitants, coupled with the lack of alternative places to live, and a strong connection to the place.

Based on these sensory impressions, we experience an ambiance that makes us feel melancholy, that hits us emotionally. We feel uncomfortable with our big cameras, feel almost voyeuristic when we take photos of Grand Isle. At the same time, we are impressed by the resilience of the island's inhabitants and have great respect for the fact that many of them still live on Grand Isle despite repeated hurricanes and damage to their homes. We can understand why they do not want to leave their island (their Mode A landscape). At the same time, we are also in complete disbelief that people keep repairing or rebuilding their houses, even though it can be assumed that the possibility of living on the island is finite. We are also particularly impressed by the fact that traditional adaptation measures to the specifics of regional Space 1 now seem exhausted and new strategies are being developed. The ever-higher elevating of houses, which now shift private life to about five meters above the ground, represent for us an unfamiliar sight—especially in view of the verticality of small towns, also unusual for us—which neither corresponds to stereotypical landscape ideas in the Mode B, nor to our respective native normal landscape (Mode A). Rather, the perception of the island takes place based on our respective Mode A of landscape, so that, through the discrepancy between what we interpret as normal and the landscape found on site, an empathic transfer of the Landscape 1a takes place.

7.3 The Beach

In general, even more so than with 'coastline', 'beach' can be understood—according to John Fiske (2003, p. 51)—as "an anomalous category between land and sea that is neither, but contains features of both". The 'beach' represents a hybrid space of land, water, and air, and is characterized by a specific quality of varying rhythmic interactions. Socially, beach can be understood as an intertwining of place and time beside the sea "outside of mundane normality", transcending work and everyday life (Fiske 2003, p. 51). While the sea is attributed to nature (Kühne, Denzer, and Eissner 2022) which is understood as "untamed, uncivilized, raw" (Fiske 2003, p. 52), the beach undergoes a social transformation into the 'anomalous category' of 'beach', associated with activities that in other physical and social contexts would be associated with the loss of social recognition (such as wearing swimwear to the office or sunbathing on a traffic island; cf. also Löfgren 2002). The 'abnormality' is also expressed in the specific furnishing of the beach by sunbeds, tables, chairs, umbrellas, etc., objects that are commonly used within closed spaces (Fiske 2003). Designated locations also provide a Space 1 and 3 opportunity for physical activities of self-experience: "Bodies perform themselves in-between direct sensation of the 'other' and various sensescapes (…) they navigate backwards and forwards between directly sensing the external world as they move bodily in and through it" (Urry and Larsen 2011, p. 196, in reference to Rodaway 2002, 1994). Beaches also have a great symbolic significance in the socialization of landscape; they are a central element of youth culture. This

Fig. 7.3 The damage to the levee catches our eye directly from the beach and illustrates the force with which the wind and water hit the island. We suspect that the sandbags used to patch the damaged levee in a makeshift fashion would not hold up for very long in the event of the next hurricane. Far less obvious is the damage to some of the houses (photos above), which we often only notice at second glance. (Photo above left, Lara Koegst 2022; above right and below, Olaf Kühne 2022)

significance arises appreciably from that of Fiske (2003, p. 68) "because youth itself is an anomalous category, that between child and adult". This meaning, in turn, is recursively solidified in the media (in song lyrics, television series, Internet videos, and films). This contributes significantly to assigning an idealized meaning to the beach and its inherent activities as a place of 'removal from everyday life' (cf. also Kiefl 2001; Kühne, Denzer, and Eissner 2022; Kühne and Weber 2019; Urbain 2003).

Against this background, the Grand Isle beach can be visually understood as an 'anomaly' of the 'abnormal category' of 'beach': Even during the day, it is largely deserted during our visit to Grand Isle, although the development beyond the levee might suggest a potential beach audience at a cursory glance. A closer look, however, shows the destruction of houses, sometimes the demolition of entire houses, leaving only their remaining foundations just above the levee—a levee that is broken in numerous places, completely missing in some places, in other places exposing its basic structure, and patched with sandbags in a makeshift manner in some places (Fig. 7.3). Another

Fig. 7.4 Impressions on the beach of Grand Isle at sunrise and sunset: In the twilight hours of the day, fishing boats are on the move near the coast in the Gulf of Mexico and, in combination with the warm light of the rising or setting sun and the silhouettes of the pelicans, create stereotypical vacation scenarios (top left and right). Accompanied by seagulls and beach walkers, our walk along the beach in the rising morning sun is very relaxing and we enjoy the view along the beach. In the process, we almost forget the destruction of the buildings; only at second glance do we notice damage to the houses, for example, in the picture shown here, the low blue/gray house in the center of the picture, whose roof has sagged forward. (Photographs: top left and bottom, Lara Koegst 2022; top right, Olaf Kühne 2022)

anomalous element is formed by an earthmover that pushes away the sand of the beach to use it for rebuilding the levee.

Nevertheless, we cannot suppress the reminiscences of vacation even on the beach of Grand Isle; after all, we taste the salt of the sea on our lips, feel the wind on our skin, walk barefoot through the sand, and hear the seagulls calling out while the waves of the sea run onto the beach with a light murmur. Turning our backs to the settlement, we find it all too easy—beyond a Mode A home-sweet-home emotional attachment—to understand why the residents of Grand Isle do not want to give up their dwellings: The view of the sea and along the long, empty beach strikes us as idyllic, almost paradisiacal. This impression is reinforced during our visits to the beach at sunset, as well as at sunrise

Fig. 7.5 View of the Gulf of Mexico from the beach of Grand Isle with oil production facilities on the horizon. In contrast to the sunset and sunrise images, the rainy weather creates a much more dramatic atmosphere. (Photo: Lara Koegst 2022)

(Fig. 7.4). We watch the pelicans, which are so typical for Louisiana sitting on the wooden pegs in the shallow water—not for nothing the pelican is the heraldic animal of Louisiana, which is also described as the 'Pelican State'. Early in the morning, in the warm light of the rising sun, we can watch fishing boats hauling in their nets near the beach. We can even, almost, overlook the oil production facilities visible at irregular intervals across the horizon, or integrate them into the ambivalent scenery of Louisiana's coastal pastiche. We remain cognitively (in a Mode C sense) aware of the risks associated with offshore oil production, we are aware of the contribution this production has made to the loss of Louisiana's coastal zone, yet it also surrounds us with the nostalgic air of the beginnings of old industrialization. Moreover, we find it difficult to imagine how the sea, whose waves rolled so calmly and steadily onto the beach during our stay at Grand Isle, can become so churned up that the water floods the entire beach, overcomes the levee (see Fig. 7.3 below), damaging and destroying and finally leaving such severe damage to the elevated houses that they have to be demolished, rebuilt, and refurbished.

Evocatively, contrasting impressions of the ambiance coalesce for us at the beach of Grand Isle: Firstly, the clearly recognizable damage caused by wind and water during the hurricane to the levee and the houses towering above it. Secondly, the quiet idyll of the beach in the most glorious sunshine, which precipitates a vacation mood in us, yet with the constant reminder of the oil drilling in the Gulf of Mexico (Fig. 7.5). Thirdly, however, the presence of the facilities of this offshore oil industry, which has contributed to the modernization of Louisiana, but also to the loss of large coastal spaces, simultaneously, challenge stereotypical expectations of beaches and are already generating emergent future memories of this industry in transition to historic relics of industrialization.

8 Conclusion

Space 1 of the Louisiana coast, with its numerous and interacting processes, is dually characterized by a high degree of complicatedness, that being a diverse multitude of existing structures, and often synchronously a high degree of complexity which arises from the different relationships between these structures and system states. From the high degree of complicatedness and complexity, in turn, a multitude of contingent states are derived. In concrete terms, this also means that it is difficult to predict where exactly which processes of land loss will be manifested. Corresponding to the subject matter, the social and scientific approach to coastal land loss is also characterized by a high degree of complicatedness, complexity, and contingency. Accordingly, a multi-perspective research approach is recommended, as we have pursued with the neopragmatic approach of horizontal geographies in this book. We will summarize its main findings again below by addressing the five extremes of coastal land loss in Louisiana that were presented in the introduction and will be revisited again in the future. Furthermore, we will once again outline the neopragmatic approach, particularly presenting the distinctions and enhancements of the chosen theoretical framework. Finally, we will address its opportunities and constraints for exploring horizontal geographies and identify key research needs.

8.1 The Five Extremes of Coastal Land Loss in Louisiana—And An Emerging Sixth

At the beginning of the introduction to this book, we described coastal Louisiana as a fivefold extreme space; this assessment was confirmed and further delineated in subsequent comments:

First, it is characterized by extreme physical dynamics. Processes of land formation and land loss intertwine. Glacial isostatic subsidence is contrasted with sedimentation of

O. Kühne and L. Koegst, *Land Loss in Louisiana*, RaumFragen: Stadt – Region – Landschaft, https://doi.org/10.1007/978-3-658-39889-7_8

the Mississippi River (now influenced by humans). The layered sediment is not static, but becomes compacted, causing displacements and the formation of salt domes (diapirism). Overall, the space—and we refer here to the period since the last glaciation—exhibits intense dynamics and complexity, including on a planetary scale.

Second, it is characterized by human interventions, some of which go to extremes. These impacts are often associated with unintended side effects. Whether intentional or unintentional, they feedback into the natural processes and hybridize to form a Landscape 1 that can be interpreted as a pastiche of a diverse cultural-natural hybridity. Such processes include the regulation of the Mississippi River's flow, which dramatically alter its sedimentation behavior. Similar processes can be seen in the drainage of land and the loading that results from construction of buildings. The unintentionally induced interactions that lead to land loss are particularly evident in the case of canal construction in the marshes: With canal construction, saltwater penetration subsequently destroys autochthonous vegetation and wave action from ships leads to additional erosion. These processes are additionally exacerbated by the rise in sea level as a result of anthropogenic climate change. Also, in terms of differentiation and intensity, the coast is an extraordinary example of human intervention in Space 1.

Third, these far-reaching dynamics from restructuring Space 1 in turn evoke dire and drastic social reactions within the local communities. The dynamics greatly exceed the adaptive capacities of the autochthonous vegetation. The severity of the restructuring of Space 1 can hardly be coped with by both locally and regionally formed incorporated cultural capital and social capital, because these are fundamentally altered, especially in the form of the shift of the status of the land and water hybridity to the pole of water. In many cases, the only option left is to migrate from the affected areas.

Fourth, these dynamics pose extreme challenges to scientific disciplines. The complexity of the diverse dynamics and interactions force an interdisciplinary approach. But not only this, as it is also unlikely to comprehensively illuminate the complexity of processes and structures from a theoretical perspective or to capture it using a single method of data collection. Also, such complexity requires special attention to the presentation of results, so as to translate complexity but neither to reduce it inappropriately—which would risk generating stereotyping—or to exhibit a level of complexity that would overwhelm readers. Moreover, such a neopragmatic redescription should make clear that it is a contingent interpretation of the development of Space/Landscape 1, 2, and 3, as well as of the relations between Spaces/Landscapes 1 and 2 as well as those of 2 and 3.

Fifth, land loss in Louisiana is associated with policy choices that tend to extremes. Primarily, this is found in over a century of support for the petrochemical industry, in the form of wide-ranging permits to encroach on Space 1 or the widespread practice of pecuniary incentives (Colten 2012; Hemmerling 2007; Hochschild 2016). Concurrently, in a fairly widespread practice of denying climate change and its consequences, which makes it difficult to implement measures of mitigation and adaptation, and thus to preserve livelihood opportunities. Additionally, a concurrent administrative decision to declare certain areas uninhabitable and exclude them from more intensive coastal protection measures (the Morganza-to-the-Gulf Hurricane Protection System levee). This mélange of political and administrative decisions tending toward extremes leads to a simultaneity of the non-simultaneous (Bloch 1962[1935]), i.e., the reconstruction or new construction of buildings in areas deemed uninhabitable in the medium term.

The analysis of the mass media response to coastal land loss, however, has shown that one extreme has so far failed to materialize; this applies to both journalistic media and social media. The intensity of the discussion about climate change in social media (Kittler 2021) could, however, also extend to land loss. All in all, however, it can be stated that this requires a more decided scientific approach, which in terms of scope exceeds a horizontal geographic investigation, in which the focus ultimately lies on the synthesis of different developments to and in Space/Landscape 1, 2 and 3, as well as the relations between Space/Landscape 1 and 2 as with Space/Landscape 2 and 3.

Concordant to the extreme nature of the developments of coastal land loss in Louisiana, also with regard to the intensity of the investigation of these processes on the part of science, conclusions from the findings obtained can also be transferred to other parts of the world that are affected with coastal land loss arising from anthropogenic climate change, and where coastal land loss has not yet progressed so far (for example, as a result of other natural conditions or lesser human interventions). In this respect, impulses for mitigation and adaptation measures can be derived. Given the extreme nature of human interventions in sensitive ecosystems, however, these often take the form of bad practice and worst practice examples. In general, these affected societies face the challenge of adapting to ongoing changes in Space 1. These adaptations are necessary to limit the loss of life chances.

8.2 The Neopragmatic Framework of Horizontal Geographies of Coastal Land Loss

Starting from the exhortation of Karl Popper (Popper 1963; 1992), which can already be found in a similar form in Max Weber's work (2010 [1904/05]) and also later by Peter Berger (Berger 2017[1963]), we have dealt with the coastal land loss in Louisiana with an intention of making the unintended side-effects of human actions the subject of social science research. Since this has natural origins as well as social ones, the challenge was to integrate findings from natural science and social science research and reflect them beyond the level of descriptiveness. To frame this integration theoretically, on the one hand, we operationalized Karl Popper's three worlds theory for spatial and landscape research by analytically separating the three levels of Space/Landscape 1 (tangible substrate), Space/Landscape 2 (individual conceptions), and Space/Landscape 3 (social conceptions, such as patterns of interpretation, valuation, and categorization). On the other hand, we have drawn on philosophical neopragmatism (particularly in the tradition of Richard Rorty) and derived from it a functionalization exploring complex objects by elaborating six levels of triangulation. This theoretical framework has proven fruitful for examining the underlying processes of varying degrees of cultural-natural hybridity in Louisiana's land loss and how they intertwine.

The neopragmatic redescription of the Louisiana coastal land loss that we have undertaken is constitutively based on the integration of theories that are often considered incommensurable, in our case especially positivist and, conversely, social constructivist and phenomenological approaches. With this combination of theories, we succeeded in examining natural scientific findings regarding coastal land loss in its social meaning and, not insignificantly, in contrasting them with our experience. Accordingly, we also used different methods of 'illuminating' the research subject, the meta-analysis of the (mostly implicitly positivist) literature, the cartographic analysis of coastal land loss, the media content analysis as well as the phenomenological walk. Especially in the media content analysis, the construction of coastal protection, in Mode A and B also became clear. In the literature analysis, this became clear when addressing the incorporation of these modes into coastal protection measures (Hemmerling et al. 2022). Regarding the cartographic representation of coastal land loss, the limitation of cartography striving for unambiguous attribution became clear: A Space 1 characterized by sporadic and rhythmic shifts between the poles of land and water hybridity is difficult to capture cartographically with a logic that adheres to any binary logic. Nevertheless, such a positivist cartography exhibits a highly functional potential of this, as it follows the system logic of Mode B politics and is thus able to generate or intensify a willingness to act in terms of mitigation and adaptation in relation to coastal land loss. However, particularly in light of the uncertainty in projections regarding the intensities of anthropogenic climate change and its consequences—in this case, the thermal expansion of the world's oceans and the meltwater of inland glacial deposits—in the future, focusing on the existing Mode B cartographic patterns of construction of coastal land loss runs the risk of excessively simplifying the inherent complexity. In turn, this is liable to trivialize corporeal ligatures in their binding effects. Different dangers are found in a hybrid Space 1 than on 'land'. This shift between the poles of land and water hybridity becomes particularly relevant in situations where areas are not only partially or temporarily submerged but also situations like the construction of the Morganza levee. Here, certain areas are defined from the outset as being 'not permanently' dry land thusly left to shift its hybrid portion in favor of water, increasing the intensity of socially mediated corporeal ligatures.

At this point, the strengths of the neopragmatic approach to horizontal geographic research become particularly apparent: Focusing on the potentials of theories, not on their boundaries of limitation, allows the testing of their core in other contexts and to extend them if necessary. This is illustrated in this context by Ralf Dahrendorf's life chances concept, here especially in its ligature component, which has been subjected to additional contrasting. The differentiation already performed by Kühne et al. (2022) into internally directed and externally directed ligatures, into moral and ethical ligatures, into implicit and explicit ligatures, could be advantageously used for our work, for example in dealing with externally directed moral ligatures in the context of the discussions about the YouTube video, but the concept—also in the mentioned extension—remains focused on the relations of World 3 and World 2 (subsequently also Space and Landscape 3 and 2). The centrality of World 1 in terms of individual and social possibilities and limits with coastal land loss remained outside the theoretical framework, although the conceptual layout of the theory of life chances (with its options and ligatures) did not exclude the inclusion of the material world, but (as a social theory) simply did not take it into account. The expansion of the concept thus represents a (possibly provisional) completion: On the one hand, to include corporeal ligatures, i.e., the restrictions to which humans are subjected as a result of their participation in World 1. On the other hand, also by corporeally mediated social ligatures, which result from the fact that other people intervene in the physical space and thus evoke certain corporeally bound actions.

Space 1, here the differentiation of Space 1, also underlies the second extension of a classical social theory. But before we turn to this differentiation, it is important to clarify another specific feature of the approach of neopragmatic horizontal geographies that we advocate: During the research process, the problem of the relationship of vulnerability and resilience on the availability of symbolic capital came to light (Birch and Carney 2020). This suggested the extension of the theoretical approach to include Pierre Bourdieu's concept of symbolic capital (Bourdieu 1984, 1985, 1989, 2002) The possibility of using this theory to make the results available for a wider field of discussion on the spatial distribution of symbolic capital justified the extension of the theoretical basis. However, results also showed the necessity of a regional specification of the concept of capital: Only in this way the Mode A of constructing world (here especially Space and Landscape) is constitutively dependent on a certain Space 1. And Mode B also shows itself to be regionally and, moreover, partially socially differentiated. Mode C, on the other hand, is largely constituted universally, but it is differentiated by the constitution of various professional discourses. It is only regionally differentiated in small areas. Mode C acquires regional relevance only when certain general structures and processes are particularly relevant regionally. The adaptation to the regional conditions of the southern Louisiana hybrid space, however, have produced specific Mode A and B social capital and incorporated cultural capital that cannot be subordinated to a universal theory of capital. In this respect, it seems necessary to us to conceptually complement this general theory of social and cultural capital with capital that is regionally specific, in other words, capital formed only regionally and equally valid. However, the notion of a Mode B shared throughout society is also becoming increasingly fragile as a result of the fragmentation and differentiation of society, especially in the U.S., while less so in Western and Central Europe, after all, there is less of a binding school canon here in the U.S. than in the 'Old Continent'. All in all, via Internet discourses, religious discourses, variously ideologically oriented television channels, social media, etc., increasingly ideologically oriented discourse-bound variants of Mode B communication are produced, so that it is increasingly questionable whether a common perception to world, space, and landscape can still be found at all or whether it is more adequate to speak of differentiated pluralities of partial societal common perceptions with certain intersections in the sense of postmodern discourse pluralities.

Not only in the scientific reference to World 3, but also World 2, post-modernization tendencies can be recognized with regard to coastal land loss. The already above discussed hybridization phenomena of water and land in Space 1, the emergence of distinct nature-culture hybrids as well as the integration into the system of urban–rural hybrids in transition indicates developments that can be described as postmodern. Their combination into a spatial pastiche that largely defies attempts to construct explicitness makes the post-modernization of southern Louisiana particularly salient. While these epistemological interpretations may represent a more capable development of vocabularies and thus have a positive effect on 'life chance' developments, it becomes clear when reflecting on the normative approach of our work: In much of coastal Louisiana, it is less a matter of maximizing life chances than minimizing their loss. Because of the multiple causes of coastal land loss, coastal Louisiana can be considered an exemplary region for the challenges facing other coastal regions around the world as climate change progresses.

8.3 Possibilities and Constraints of the Neopragmatic Approach of Horizontal Geographies

With the neopragmatic approach of horizontal geographies, a challenge arises that is also faced by other theories that aim at overcoming the subject-object duality, such as actor-network theory and assemblage theory (see in spatial sciences, for instance: Anderson 2015; Färber 2014; Mattissek and Wiertz 2014; Miggelbrink 2014; Müller 2015) or those which are conceived as a 'grand theory', like the systems theory of Niklas Luhmann (among many: Kneer and Nassehi 1997; Luhmann 1986, 1987, 1990). The development not only of a specific way of thinking with which worlds are to be penetrated, but, because language also has a performative effect, also the development of its own terminology. This, in turn, makes it difficult for even theory-sensitive readers who have not (yet) studied the theoretical framework in depth to gain immediate access. In this respect, when applying a neopragmatic theoretical framework of horizontal geographic research, there is the coexisting challenge of justifying the selection and application of theories along with the demand for transdisciplinarity, because interdisciplinary and transdisciplinary are not only relevant in the collection of data and the interpretation of (preliminary) results, but also in the communication of neopragmatic redescriptions. Accordingly, a special focus has to be put on the presentation of the results in a way that they are comprehensible to persons with differing Mode C contexts and can also be understood by persons who have a Mode B approach to world in this subject area.

We take these challenges as an opportunity to further develop the theory and practice of neopragmatic horizontal geography. The great effort of justification regarding the decisions for the choice of certain theories, methods, collections of data, researcher perspectives, the inclusion of different modes of world construction as well as representations—measured by the results achieved—can be described as justified, not least because pivotal approaches are thus laid open to an intersubjective comprehension. This facilitates a factual critique of the individual decisions, which in turn provides an opportunity to further develop the approach. The critical appraisal of the approach is not limited to an external input, but a constant reflection on the appropriateness of the choice of theory in the context of the results obtained also allows for a revision or enhancement of the theoretical framework used, in our case the addition of Pierre Bourdieu's theory of capital, which was not initially foreseen. Also, this circular research process, the constant reconnection of theory, methodology, acquired data bases, involvement of third parties, as well as the (graphical) presentation of results not only allows for a high flexibility in the choice of theory and methodology, but also for the targeted differentiation and expansion of the theories used, in this case Bourdieu's theory of capital and Ralf Dahrendorf's life chances approach.

Overall, it can be stated that the presented neopragmatic redescription of land loss has hitherto scarcely related disciplinary research traditions to each other, and in some places has furthermore achieved and integrated its own empirical results. However, research gaps also became clear. These particularly relate to the study of mass media communication, especially in social media, of land loss. In our exploratory study, we were able to highlight some aspects of this communicative construction of world, but our effort is far from a systematic coverage. Since it is precisely this form of world construction that is of increasing importance in the emergence and dissemination of patterns of interpretation, evaluation, and categorization, a more differentiated scientific approach seems warranted here. This also applies to the intercultural comparison of the social mass media communication of catastrophes and hazards, on which we were able to achieve initial results. The starting point of our reflections on neopragmatic redescriptions of horizontal geographies of coastal land loss was its observation from a natural science (positivist) perspective; this complements other approaches of neopragmatic redescriptions that started from social constructions of space (such as: Kühne and Jenal 2020a; Kühne and Weber 2019). In this respect, from a metatheoretical perspective—in terms of determining the limits of the neopragmatic approach's suitability—it also seems worthwhile to use other theories as starting theories, such as phenomenology or critical theory. Such an approach is also done with the awareness of providing a contingent redescription of world.

References

Ackermann, U. (2020). *Das Schweigen der Mitte. Wege aus der Polarisierungsfalle.* Darmstadt: WBG.

Aitken, S. C., & Craine, J. W. (2015). A Brief History of Mediated, Sensational and Virtual Geographies. In S. P. Mains, J. Cupples, & C. Lukinbeal (Eds.), *Mediated Geographies and Geographies of Media* (pp. 81–92). Dordrecht: Springer.

Alt, J. A. (1995). *Karl R. Popper* (2nd ed.). Frankfurt: Campus-Verlag.

Althoff, T., White, R. W., & Horvitz, E. (2016). Influence of Pokémon Go on Physical Activity: Study and Implications. *Journal of Medical Internet Research, 17,* (*12,* 1–14).

American Institute of Professional Geologists. (1993). *A Citizen's Guide to Geologic Hazards. A Guide to Understanding Geologic Hazards Including Asbestos, Radon, Swelling Soils, Earthquakes, Volcanoes.* Arvada, Colorado: Self-Publishing.

Anderson, J. (2015). Towards an Assemblage Approach to Literary Geography. *Literary Geographies, 1,* (*2,* 120–137).

Arnberger, E. (1993). *Thematische Kartographie* (3rd ed.). Braunschweig: Westermann.

Aschenbrand, E. (2016). Einsamkeit im Paradies. Touristische Distinktionspraktiken bei der Aneignung von Landschaft. *Berichte. Geographie und Landeskunde, 90,* (*3.* 219–234).

Bahr, H.-D. (2014). *Landschaft. Das Freie und seine Horizonte.* Freiburg: Alber Verlag.

Baltzer, U. (2001). Rorty und die Erneuerung des Pragmatismus. In T. Tietz, & U. Schäfer (Eds.), *Hinter den Spiegeln. Beiträge zur Philosophie Richard Rortys mit Erwiderungen von Richard Rorty* (pp. 21–48). Frankfurt am Main: Suhrkamp.

Barnes, T. J. (2008). American pragmatism: Towards a geographical introduction. *Geoforum, 39,* (*4,* 1542–1554). https://doi.org/10.1016/j.geoforum.2007.02.013.

Barras, J., Beville, S., Britsch, D., Hartley, S., Hawes, S., Johnston, J., et al. (2004). Historical and Projected Coastal Louisiana Land Changes: 1978–2050. USGS Open-File Report. OFR 03-334 (Revised January 2004), U.S. Department of the Interior; U.S. Geological Survey. https://pubs.er.usgs.gov/publication/ofr03334. Accessed: 2 December 2021.

Barras, J. A. (2007). Land Area Changes in Coastal Louisiana. After Hurricanes Katrina and Rita. In G. S. Farris, G. J. Smith, M. P. Crane, C. R. Demas, L. L. Robbins, & D. Lavoie (Eds.), *Science and the storms. The USGS Response to the Hurricanes of 2005* (pp. 97–102). Reston, Virginia: Government Printing Office.

Barras, J. A. (2009). *Land Area Change and Overview of Major Hurricane. Impacts in Coastal Louisiana, 2004–08.* Reston, Virginia: U.S. Department of the Interior.

Barry, J. M. (1998). *Rising Tide. The Great Mississippi Flood of 1927 and How It Changed America.* Riverside: Simon & Schuster.

O. Kühne and L. Koegst, *Land Loss in Louisiana*, RaumFragen: Stadt – Region – Landschaft, https://doi.org/10.1007/978-3-658-39889-7

Bass, A. S., & Turner, R. E. (1997). Relationships between Salt Marsh Loss and Dredged Canals in Three Louisiana Estuaries. *Journal of Coastal Research, 13,* (*3.* 895–903). https://www.jstor.org/stable/4298682.

Baum, L. (2021). Umweltveränderungen, Klimawandel und sozialräumliche Folgen auf der Isle de Jean Charles—Eine kritische Analyse des medialen Diskurses. In O. Kühne, T. Sedelmeier, & C. Jenal (Eds.), *Louisiana—mediengeographische Beiträge zu einer neopragmatischen Regionalen Geographie* (pp. 151–166). Wiesbaden: Springer.

Bauman, Z. (1990). Modernity and Ambivalence. *Theory, Culture & Society, 7,* (*2–3,* 143–169).

Bauman, Z. (2010). *Liquid Times. Living in an Age of Uncertainty.* Cambridge: Polity Press.

Beatley, T. (2009). *Planning for Coastal Resilience: Best Practices for Calamitous Times.* Washington, D.C.: Island Press.

Beck, U. (1986). *Risikogesellschaft. Auf dem Weg in eine andere Moderne.* Frankfurt (Main): Suhrkamp.

Beck, U. (2006). *Weltrisikogesellschaft. Auf der Suche nach der verlorenen Sicherheit.* Frankfurt (Main).

Belina, B., & Miggelbrink, J. (Eds.). (2010). *Hier so, dort anders. Raumbezogene Vergleiche in der Wissenschaft und anderswo* (Raumproduktionen, vol. 6, 1. Auflage). Münster: Westfälisches Dampfboot.

Berger, P. L. (2017[1963]. *Einladung zur Soziologie. Eine humanistische Perspektive* (UTB Soziologie, vol. 3495, 2., ergänzte Auflage). Konstanz: UVK Verlagsgesellschaft mbH; UVK/Lucius (Original work published 1963).

Bernier, J. (U.S. Geological Survey, Ed.). (2013). Trends and Causes of Historical Wetland Loss in Coastal Louisiana, US Department of the Interior. https://pubs.usgs.gov/fs/2013/3017/. Accessed: 18 February 2021.

Bernstein, F., Kaußen, L., & Stemmer, B. (2019). Online-Partizipation und Landschaft. In O. Kühne, F. Weber, K. Berr, & C. Jenal (Eds.), *Handbuch Landschaft* (pp. 547–558). Wiesbaden: Springer VS.

Berr, K. (2017). Zur Moral des Bauens, Wohnens und Gebauten. In K. Berr (Ed.), *Architektur- und Planungsethik. Zugänge, Perspektiven, Standpunkte* (pp. 111–138). Wiesbaden: Springer VS.

Berr, K., Jenal, C., Koegst, L., & Kühne, O. (2022). *Noch mehr Sand im Getriebe? Kommunikations- und Interaktionsprozesse zwischen Landes- und Regionalplanung, Politik und Unternehmen der Gesteinsindustrie* (RaumFragen—Stadt—Region—Landschaft). Wiesbaden: VS Springer.

Birch, T., & Carney, J. (2020). Regional Resilience: Building Adaptive Capacity and Community Well-Being Across Louisiana's Dynamic Coastal–Inland Continuum. In S. Laska (Ed.), *Louisiana's Response to Extreme Weather. A Coastal State's Adaptation Challenges and Successes* (Springer eBooks Earth and Environmental Science, 1st ed., pp. 313–340). Cham: Springer.

Bischoff, W. (2005). *Nicht-visuelle Dimensionen des Städtischen: olfaktorische Wahrnehmung in Frankfurt am Main, dargestellt an zwei Einzelstudien zum Frankfurter Westend und Ostend.* (Dissertation, Johann Wolfgang Goethe-Universität). Frankfurt (Main).

Bischoff, W. (2007). "Korrespondierende Orte"—Zum Erscheinen olfaktorischer Stadtlandschaften. In C. Berndt, & R. Pütz (Eds.), *Kulturelle Geographien. Zur Beschäftigung mit Raum und Ort nach dem Cultural Turn* (pp. 189–212). Bielefeld: transcript.

Bishop, J. (2014). Representations of 'trolls' in mass media communication: a review of media-texts and moral panics relating to 'internet trolling'. *International Journal of Web Based Communities, 10,* (*1,* 7–24). https://doi.org/10.1504/IJWBC.2014.058384.

Bisschop, L. C. J., Strobl, S., & Viollaz, J. S. (2018). Getting into Deep Water: coastal land loss and State-Corporate Crime in the Louisiana Bayou. *The British Journal of Criminology, 58,* (*4,* 886–905). https://doi.org/10.1093/bjc/azx057.

Blackbourn, D. (2007). *Die Eroberung der Natur. Eine Geschichte der deutschen Landschaft.* München: Random House.

Bloch, E. (1962[1935]). *Erbschaft dieser Zeit* (Gesamtausgabe, Band 4, Erweiterte Ausgabe). Frankfurt am Main: Suhrkamp.

Boesch, D. F. (2020). Managing Risks in Louisiana's Rapidly Changing Coastal Zone. In S. Laska (Ed.), *Louisiana´s Response to Extreme Weather. A Coastal State's Adaptation Challenges and Successes* (pp. 35–62). Berlin, Heidelberg: Springer Nature.

Boesch, D. F., Josselyn, M. N., Mehta, A. J., Morris, J. T., Nuttle, W. K., Simenstad, C., et al. (1994). Scientific Assessment of Coastal Wetland Loss, Restoration and Management in Louisiana. *Journal of Coastal Research, (Special Issue 20,* 1–103).

Bork-Hüffer, T., Füller, H., & Straube, T. (Eds.). (2021). *Handbuch Digitale Geographien. Welt—Wissen—Werkzeuge.* Paderborn: Brill Schöningh.

Böse, H., Stemmer, B., Moczek, N., & Hofmann, M. (2019). Die Bedeutung der Ortsidentität für die Landschaftswahrnehmung am Beispiel von Windenergieanlagen. In M. Hülz, O. Kühne, & F. Weber (Eds.), *Heimat. Ein vielfältiges Konstrukt* (pp. 179–202). Wiesbaden: Springer VS.

Bourassa, S. C. (1991). *The Aesthetics of Landscape.* London: Belhaven Press.

Bourdieu, P. (1984). *Distinction. A Social Critique of the Judgement of Taste.* Cambridge MA: Harvard University Press.

Bourdieu, P. (1985). The Market of Symbolic Goods. *Poetics, 14, (1–2,* 13–44).

Bourdieu, P. (1989). Social space and symbolic power. *Sociological theory, 7, (1,* 14–25).

Bourdieu, P. (2002). Das ökonomische Feld. In P. Bourdieu (Ed.), *Der Einzige und sein Eigenheim* (Schriften zu Politik & Kultur, vol. 3, pp. 185–222). Hamburg: VSA.

Britsch, L. D., & Cunbar, J. B. (1993). Land Loss Rates: Louisiana Coastal Plain. *Journal of Coastal Research, 9, (2,* 324–338). https://doi.org/10.5724/gcs.91.12.0034.

Britsch, L. D., & Kemp, E. B. (1991). Land Loss Rates: Louisiana Coastal Plain. In K. W. Shanley, & B. F. Perkins (Eds.), *Coastal Depositional Systems in the Gulf of Mexico. Quaternary Framework and Environmental Issues.* SEPM Society for Sedimentary Geology.

Brunnengräber, A. (2018). Klimaskeptiker im Aufwind. Wie aus einem Rand- ein breiteres Gesellschaftsphänomen wird. In O. Kühne, & F. Weber (Eds.), *Bausteine der Energiewende* (pp. 271–292). Wiesbaden: Springer VS.

Bruns, D., Kühne, O., Schönwald, A., & Theile, S. (Eds.). (2015). *Landscape Culture—Culturing Landscapes. The Differentiated Construction of Landscapes.* Wiesbaden: Springer VS.

Brüntrup, G. (1996). *Das Leib-Seele-Problem. Eine Einführung.* Stuttgart: Kohlhammer.

Bunge, M. (1984). *Das Leib-Seele-Problem. Ein psychobiologischer Versuch* (Die Einheit der Gesellschaftswissenschaften, vol. 37). Tübingen: J.C.B. Mohr (Paul Siebeck).

Burckhardt, L. (2006). Spaziergangswissenschaft (1995). In M. Ritter, & M. Schmitz (Eds.), *Warum ist Landschaft schön? Die Spaziergangswissenschaft* (pp. 257–300). Kassel: Martin Schmitz Verlag.

Burgess, J., & Gold, J. R. (Eds.). (1985). *Geography The Media & Popular Culture.* New York: St. Martin's Press.

Burgess, J., & Green, J. (2018). *YouTube. Online Video and Participatory Culture* (Digital Media and Society Series, 2nd ed.). Cambridge: Polity Press.

Burgess, J. A. (1985). News from Nowhere: The Press, the Riots and the Myth of the Inner City. In J. Burgess, & J. R. Gold (Eds.), *Geography The Media & Popular Culture* (pp. 192–228). New York: St. Martin's Press.

Cahoon, D. R., Reed, D. J., Day, J. W., Steyer, G. D., Boumans, R. M., Lynch, J. C., et al. (1995). The Influence of Hurricane Andrew on Sediment Distribution in Louisiana Coastal Marshes. *Journal of Coastal Research, (Special Issue 21,* 280–294).

Carey, M. (1814). Louisiana. In M. Carey (Ed.), *Carey's General Atals, Improved And Enlarged. Being A Collection Of Maps Of The World And Quarters, Their Principal Empires, Kingdoms, &c.* (p. 26). Philadelphia: T. S. Manning, Printer, N. W. Corner of Sixth & Chesnut Streets. https://www.davidrumsey.com/luna/servlet/detail/RUMSEY~8~1~635~50074:Louisiana-?title=Search+Results%3A+List_No+equal+to+%274577.026%27&thumbnailViewUrlKey=link.view.search.url&fullTextSearchChecked=&dateRangeSearchChecked=&showShareIIIFLink=true&helpUrl=https%3A%2F%2Fdoc.lunaimaging.com%2Fdisplay%2FV75D%2FLUNA%2BViewer%23LUNAViewer-LUNAViewer&showTip=false&showTipAdvancedSearch=false&advancedSearchUrl=https%3A%2F%2Fdoc.lunaimaging.com%2Fdisplay%2FV75D%2FSearching%23Searching-Searching. Accessed: 4 July 2022.

Chilla, T., Kühne, O., Weber, F., & Weber, F. (2015). "Neopragmatische" Argumente zur Vereinbarkeit von konzeptioneller Diskussion und Praxis der Regionalentwicklung. In O. Kühne, & F. Weber (Eds.), *Bausteine der Regionalentwicklung* (pp. 13–24). Wiesbaden: Springer VS.

Clipp, A., Gentile, B., Green, M., Galinski, A., Harlan, R., Rosen, Z., et al. (2017). Coastal Master Plan: Appendix B: People and the landscape. Version Final, Coastal Protection and Restoration Authority. http://coastal.la.gov/wp-content/uploads/2017/04/Appendix-B_People-and-the-Landscape_FINAL.pdf. Accessed: 31 May 2022.

Coastal Protection and Restoration Authority of Louisiana. (2017). Louisiana's Comprehensive Master Plan for a Sustainable Coast. Effective June 2, 2017. http://coastal.la.gov/wp-content/uploads/2017/04/2017-Coastal-Master-Plan_Web-Book_CFinal-with-Effective-Date-06092017.pdf. Accessed: 24 August 2021.

Colten, C. E. (2006). *An Unnatural Metropolis. Wresting New Orleans From Nature.* Baton Rouge: Louisiana State University Press.

Colten, C. E. (2012). An Incomplete Solution: Oil and Water in Louisiana. *The Journal of American History, 99,* (*1,* 91–99). https://doi.org/10.1093/jahist/jas023.

Colten, C. E. (2015). The place for humans in Louisiana coastal restoration. *Labor & Engenho, 9,* (*4,* 6–18).

Colten, C. E. (2017). Environmental Management in Coastal Louisiana: A Historical Review. *Journal of Coastal Research, 33,* (*3,* 699–711). https://doi.org/10.2112/JCOASTRES-D-16-00008.1.

Colten, C. E. (2018). Cartographic Depictions of Louisiana Land Loss: A Tool for Sustainable Policies. *Sustainability, 10,* (*3,* 763). https://doi.org/10.3390/su10030763.

Colten, C. E. (2021a). Redirecting sediment and rearranging social justice. *Water History, 13,* (*1,* 33–43). https://doi.org/10.1007/s12685-020-00255-3.

Colten, C. E. (2021b). *State of disaster. A historical geography of Louisiana's land loss crisis.* Baton Rouge: Louisiana State University Press.

Colten, C. E., & Giancarlo, A. (2011). Losing Resilience on the Gulf Coast: Hurricanes and Social Memory. *Environment: Science and Policy for Sustainable Development, 53,* (*4,* 6–19).

Colten, C. E., Simms, J. R. Z., Grismore, A. A., & Hemmerling, S. A. (2018). Social justice and mobility in coastal Louisiana, USA. *Regional Environmental Change, 18,* (*2,* 371–383). https://doi.org/10.1007/s10113-017-1115-7.

Couvillion, B. R., Barras, J. A., Steyer, G. D., Sleavin, W., Fischer, M., Beck, H., et al. (2011). Land Area Change in Coastal Louisiana from 1932 to 2010. https://pubs.usgs.gov/sim/3164/downloads/SIM3164_Pamphlet.pdf. Accessed: 24 January 2020.

Couvillion, B. R., Steyer, G. D., Wang, H., Beck, H. J., & Rybczyk, J. M. (2013). Forecasting the Effects of Coastal Protection and Restoration Projects on Wetland Morphology in Coastal Louisiana under Multiple Environmental Uncertainty Scenarios. *Journal of Coastal Research, 67,* (*67 (10067),* 29–50). https://doi.org/10.2307/23486535.

Couvillion, B. R., Beck, H., Schoolmaster, D., & Fischer, M. (2017). Land area change in coastal Louisiana (1932 to 2016). https://pubs.usgs.gov/sim/3381/sim3381_pamphlet.pdf. Accessed: 24 January 2020.

Craig, N. J., Turner, R. E., & Day, J. W. (1979a). Land loss in coastal Louisiana (U.S.A.). *Environmental Management, 3,* (*2,* 133–144). https://doi.org/10.1007/BF01867025.

Craig, N. J., Turner, R. E., & Day, J. W. (1979b). Land Loss in Coastal Louisiana (U.S.A.). *Environmental Management, 3,* (*2,* 133–144).

Crampton, J., & Krygier, J. (2005). An Introduction to Critical Cartography. *ACME: An International Journal for Critical Geographies, 4,* (*1,* 11–33).

Crepelle, A. (2018). The United States first climate relocation: recognition, relocation, and indigenous rights at the Isle de Jean Charles. *Belmont Law Review, 6,* (*1,* 1–40).

Currid-Halkett, E. (2021). *Fair gehandelt? Wie unser Konsumverhalten die Gesellschaft spaltet.* München: btb.

Cutchin, M. P. (2008). John Dewey's metaphysical ground-map and its implications for geographical inquiry. *Geoforum, 39,* (*4,* 1555–1569). https://doi.org/10.1016/j.geoforum.2007.01.014.

Dahrendorf, R. (1963). *Die angewandte Aufklärung. Gesellschaft und Soziologie in Amerika.* München: Piper.

Dahrendorf, R. (1968). *Pfade aus Utopia. Arbeiten zur Theorie und Methode der Soziologie.* München: Piper.

Dahrendorf, R. (1979). *Lebenschancen. Anläufe zur sozialen und politischen Theorie* (Suhrkamp-Taschenbuch, vol. 559). Frankfurt (Main): Suhrkamp.

Dahrendorf, R. (1983). *Die Chancen der Krise. Über die Zukunft des Liberalismus.* Stuttgart: Deutsche Verlags-Anstalt DVA.

Dahrendorf, R. (1992). *Der moderne soziale Konflikt. Essay zur Politik der Freiheit.* Stuttgart: Deutsche Verlags-Anstalt DVA.

Dahrendorf, R. (2002). *Über Grenzen. Lebenserinnerungen.* München: C.H. Beck.

Dahrendorf, R. (2006). *Versuchungen der Unfreiheit. Die Intellektuellen in Zeiten der Prüfung.* München: C.H. Beck.

Dahrendorf, R. (2007). *Auf der Suche nach einer neuen Ordnung. Vorlesungen zur Politik der Freiheit im 21. Jahrhundert* (Krupp-Vorlesungen zu Politik und Geschichte am Kulturwissenschaftlichen Institut im Wissenschaftszentrum Nordrhein-Westfalen, vol. 3, 4. Auflage). München: C.H. Beck.

Dajko, N. (2020). *French on Shifting Ground. Cultural and Coastal Erosion in South Louisiana.* Jackson: University Press of Mississippi.

Deffner, V., & Haferburg, C. (2014). Pierre Bourdieu: Habitus und Habitat als Verhältnis von Subjekt, Sozialem und Macht. In J. Oßenbrügge, & A. Vogelpohl (Eds.), *Theorien in der Raum- und Stadtforschung. Einführungen* (pp. 328–347). Münster: Westfälisches Dampfboot.

Deines, S. (2008). *Situierte Kritik. Modelle kritischer Praxis in Hermeneutik, Poststrukturalismus und Neopragmatismus.* Frankfurt a. M.: Johann Wolfgang Goethe-Universität.

Denzer, V., Köppe, H., Sachs, K., & Kühne, O. (2010). Stadtstrände—Urlaubsoasen im urbanen Raum? Erste empirische Annäherungen—ein Werkstattbericht. In K. Wöhler, A. Pott, & V. Denzer (Eds.), *Tourismusräume. Zur soziokulturellen Konstruktion eines globalen Phänomens* (1st ed., 191–106). Bielefeld: transcript.

Denzin, N. K. (2007). Triangulation. In G. Ritzer (Ed.), *The Blackwell Encyclopedia of Sociology.* Oxford, UK: John Wiley & Sons.

Dewey, J. (1996). *Die Öffentlichkeit und ihre Probleme.* Berlin: Philo Verlagsgesellschaft.

Dewey, J. (2016). *Logik. Die Theorie der Forschung* (Suhrkamp Taschenbuch Wissenschaft, 2. Auflage). Frankfurt am Main: Suhrkamp.

Diaconu, M. (2010). Vom Treiben. Dérive als Methode. *Paragrana, 18,* (*2,* 121–137). https://doi.org/10.1524/para.2009.0030.

Dickmann, F. (2018). *Kartographie* (Das Geographische Seminar). Braunschweig: Westermann.

Dodge, M., McDerby, M., & Turner, M. (Eds.). (2008). *Geographic Visualization. Concepts, Tools and Applications.* Chichester, West Sussex: John Wiley & Sons, Ltd.

Dokka, R. K. (2006). Modern-day tectonic subsidence in coastal Louisiana. *Geology, 34,* (*4,* 281–284). https://doi.org/10.1130/G22264.1.

Dokka, R. K. (2011). The role of deep processes in late 20th century subsidence of New Orleans and coastal areas of southern Louisiana and Mississippi. *Journal of Geophysical Research, 116,* (*B6,* 1–25). https://doi.org/10.1029/2010JB008008.

Dörfler, T., & Rothfuß, E. (2018). Lebenswelt, Leiblichkeit und Resonanz: Eine raumphänomenologisch-rekonstruktive Perspektive auf Geographien der Alltäglichkeit. *Geographica Helvetica, 73,* (*1,* 95–107). https://doi.org/10.5194/gh-73-95-2018.

Döring, M., Stettekorn, W., & Storch, H. v. (Eds.). (2005). *Küstenbilder, Bilder der Küste. Interdisziplinäre Ansichten, Ansätze und Konzepte.* Hamburg: Hamburg University Press.

Döring, J., & Thielmann, T. (2009). Mediengeographie: Für eine Geomedienwissenschaft. In J. Döring, & T. Thielmann (Eds.), *Mediengeographie. Theorie—Analyse—Diskussion* (pp. 9–64). Bielefeld: transcript.

Drexler, D. (2013). Landscape, Paysage, Landschaft, Táj. The Cultural Background of Landscape Perceptions in England, France, Germany, and Hungary. *Journal of Ecological Anthropology, 16,* (*1,* 85–96). https://doi.org/10.5038/2162-4593.16.1.7.

Dunkel, A. (2015). Visualizing the perceived environment using crowdsourced photo geodata. *Landscape and Urban Planning, 142,* (173–186).

Eckardt, F. (2014). *Stadtforschung. Gegenstand und Methoden.* Wiesbaden: Springer VS.

Edler, D., & Dickmann, F. (2019). Landschaft im amtlichen Geoinformationswesen. In O. Kühne, F. Weber, K. Berr, & C. Jenal (Eds.), *Handbuch Landschaft* (pp. 507–515). Wiesbaden: Springer VS.

Edler, D., & Kühne, O. (2019). Nicht-visuelle Landschaften. In O. Kühne, F. Weber, K. Berr, & C. Jenal (Eds.), *Handbuch Landschaft* (pp. 599–612). Wiesbaden: Springer VS.

Edler, D., & Kühne, O. (2022). Deviant Cartographies: A Contribution to Post-critical Cartography. *KN—Journal of Cartography and Geographic Information,* (1–14). https://doi.org/10.1007/s42489-022-00110-w.

Edler, D., Kühne, O., Jenal, C., Vetter, M., & Dickmann, F. (2018). Potenziale der Raumvisualisierung in Virtual Reality (VR) für die sozialkonstruktivistische Landschaftsforschung. *KN—Journal of Cartography and Geographic Information, 68,* (*5,* 245–254). https://doi.org/10.1007/BF03545421.

Edler, D., Kühne, O., Keil, J., & Dickmann, F. (2019). Audiovisual Cartography: Established and New Multimedia Approaches to Represent Soundscapes. *KN—Journal of Cartography and Geographic Information, 69,* (5–17). https://doi.org/10.1007/s42489-019-00004-4.

Edler, D., Keil, J., Wiedenlübbert, T., Sossna, M., Kühne, O., & Dickmann, F. (2019b). Immersive VR Experience of Redeveloped Post-industrial Sites: The Example of "Zeche Holland" in Bochum-Wattenscheid. *KN—Journal of Cartography and Geographic Information, 38,* (*3,* 1–18). https://doi.org/10.1007/s42489-019-00030-2.

Ellmers, L. (2019). Politische Geographie und Landschaft. In O. Kühne, F. Weber, K. Berr, & C. Jenal (Eds.), *Handbuch Landschaft* (pp. 397–406). Wiesbaden: Springer VS.

Endreß, S. (2021). Beton, Parfüm, Fastfood—Geruchslandschaften. Phänomenologische Forschungsergebnisse eines Smellwalkes. *Stadt+Grün,* (*6,* 25–31).

Escher, A. (2006). The Geography of Cinema—A Cinematic World. *Erdkunde, 60,* (*4,* 307–314).

Escher, A., & Zimmermann, S. (2001). Geography meets Hollywood. Die Rolle der Landschaft im Spielfilm. *Geographische Zeitschrift, 89,* (*4,* 227–236).

Färber, A. (2014). Potenziale freisetzen: Akteur-Netzwerk-Theorie und Assemblageforschung in der interdisziplinären kritischen Stadtforschung. *sub\urban—zeitschrift für kritische stadtforschung, 2,* (*1,* 95–103).

Fearnley, S. M. (2009). Hurricane impact and recovery shoreline change analysis of the Chandeleur Islands, Louisiana, USA: 1855 to 2005. *Geo-Marine Letters, 29,* (*6,* 455–466).

Felgenhauer, T. (2011). Geographies of Infrastructure Systems: The Users's Lifeworld and Interface Design. *Tijdschrift voor Economische en Sociale Geografie, 103,* (*4,* 385–395). https://doi.org/10.1111/j.1467-9663.2011.00691.x.

Feyerabend, P. (2010 [1975]). *Against Method: Outline of an Anarchist Theory of Knowledge.* London: Verso.

Field, K. (2018). *Cartography. A compendium of design thinking for mapmakers* (1st ed.). Redlands, California: ESRI Press.

Fiske, J. (2003). *Lesarten des Populären* (Cultural studies, vol. 1). Wien: Löcker.

Flick, U. (2007). Triangulation in der qualitativen Forschung. In U. Flick, E. v. Kardorff, & I. Steinke (Eds.), *Qualitative Forschung. Ein Handbuch* (pp. 309–318). Reinbek bei Hamburg: Rowohlt.

Flick, U. (2011). *Triangulation.* Wiesbaden: Springer Fachmedien.

Flora Annie Steel. (1922). *The Story of the Three Little Pigs.*

Fontaine, D. (2017). *Simulierte Landschaften in der Postmoderne. Reflexionen und Befunde zu Disneyland, Wolfersheim und GTA V.* Wiesbaden: Springer VS.

Fontaine, D. (2019). Landschaft in Schulbüchern. In O. Kühne, F. Weber, K. Berr, & C. Jenal (Eds.), *Handbuch Landschaft* (pp. 641–650). Wiesbaden: Springer VS.

Fontaine, D. (2020a). Landscape in Computer Games—The Examples of GTA V and Watch Dogs 2. In D. Edler, C. Jenal, & O. Kühne (Eds.), *Modern Approaches to the Visualization of Landscapes* (pp. 293–306). Wiesbaden: Springer VS.

Fontaine, D. (2020b). Virtuality and Landscape. In D. Edler, C. Jenal, & O. Kühne (Eds.), *Modern Approaches to the Visualization of Landscapes* (267–278). Wiesbaden: Springer VS.

Fontaine, D. (2021). Wald im Schulbuch. In K. Berr, & C. Jenal (Eds.), *Wald in der Vielfalt möglicher Perspektiven. Von der Pluralität lebensweltlicher Bezüge und wissenschaftlichen Thematisierungen* (pp. 395–411). Wiesbaden: Springer VS.

Foucault, M. (1990). Andere Räume. In K. Barck, P. Gente, & H. Paris (Eds.), *Aisthesis. Wahrnehmung heute oder Perspektiven einer anderen Ästhetik* (pp. 34–46). Leipzig: Reclam.

Foxley, A. (2010). *Distance & Engagement. Walking, Thinking and Making Landscape.* Baden: Lars Müller Publishers.

Franco, G. (Ed.). (2019). *Handbuch Karl Popper* (Living reference work). Wiesbaden: Springer Reference Geisteswissenschaften.

Frazier, B. (2006). The ethics of Rortian redescription. *Philosophy & Social Criticism, 32,* (*4,* 461–492). https://doi.org/10.1177/0191453706064020.

Fröhlich, H. (2007). *Das neue Bild der Stadt. Filmische Stadtbilder und alltagliche Raumvorstellungen im Dialog* (Erdkundliches Wissen, vol. 142). Stuttgart: Franz Steiner Verlag.

Furia, P. (2021). Space and Place. A Morphological Perspective. *Axiomathes,* (1–18). https://doi.org/10.1007/s10516-021-09539-6.

Geimer, A. (2019). YouTube-Videos und ihre Genres als Gegenstand der Filmsoziologie. In A. Geimer, C. Heinze, & R. Winter (Eds.), *Handbuch Filmsoziologie* (Springer eBook Collection, pp. 1–14). Wiesbaden: Springer VS.

Gethmann, C. F. (1987). Vom Bewusstsein zum Handeln. Pragmatische Tendenzen in der deutschen Philospohie der ersten Jahrzehnte des 20. Jahrhunderts. In H. Stachowiak (Ed.), *Pragmatik. Handbuch pragmatisches Denken.* Band 2 (pp. 202–232). Leipzig: Meiner.

Glick, P., Clough, J., Polaczyk, A., Couvillion, B., & Nunley, B. (2013). Potential Effects of Sea-Level Rise on Coastal Wetlands in Southeastern Louisiana. *Journal of Coastal Research, (63 (10063),* 211–233). https://doi.org/10.2112/SI63-0017.1.

Gloy, K. (2004). *Wahrheitstheorien. Eine Einführung.* Tübingen: Francke.

Google Maps. (2022). Louisiana. https://www.google.com/maps/place/Louisiana,+USA/@29.3843316,-90.6895053,9z/data=!4m5!3m4!1s0x8620a454b2118265:0xc955f73281e54703!8m2!3d30.9842977!4d-91.9623327. Accessed: 15 May 2022.

Gotham, K. F. (2016). Coastal Restoration as Contested Terrain: Climate Change and the Political Economy of Risk Reduction in Louisiana. *Sociological Forum, 31,* (787–806). https://doi.org/10.1111/socf.12273.

Grau, A. (2017). *Hypermoral. Die neue Lust an der Empörung* (2. Auflage). München: Claudius.

Grau, A. (2019). Säkularisierung und Selbsterlösung. Die identitätslinke Läuterungsagenda als Religionsderivat. In S. Kostner (Ed.), *Identitätslinke Läuterungsagenda. Eine Debatte zu ihren Folgen für Migrationsgesellschaften* (pp. 143–150). Stuttgart: Ibidem-Verlag.

Großheim, M. (2010). Von der Maigret-Kultur zur Sherlock Holmes-Kultur. Oder: Der phänomenologische Situationsbegriff als Grundlage einer Kulturkritik. In M. Großheim, & S. Kluck (Eds.), *Phänomenologie und Kulturkritik. Über die Grenzen der Quantifizierung* (Neue Phänomenologie, vol. 15, pp. 52–84). Freiburg: Karl Alber.

Großheim, M., & Kluck, S. (2010). Phänomenologie und Kulturkritik. Eine Annäherung. In M. Großheim, & S. Kluck (Eds.), *Phänomenologie und Kulturkritik. Über die Grenzen der Quantifizierung* (Neue Phänomenologie, vol. 15, pp. 9–36). Freiburg: Karl Alber.

Gryl, I. (2022). Spaces, Landscapes and Games: the Case of (Geography) Education using the Example of Spatial Citizenship and Education for Innovativeness. In D. Edler, O. Kühne, & C. Jenal (Eds.), *The Social Construction of Landscape in Games* (pp. 359–376). Wiesbaden: Springer.

Gryl, I., Nehrdich, T., & Vogler, R. (Eds.). (2013). *geo@web. Medium, Räumlichkeit und geographische Bildung.* Wiesbaden: Springer VS.

Gustafsson, M., & Brandom, R. B. (2001). Rorty and His Critics. *The Philosophical Review, 110,* (*4,* 645). https://doi.org/10.2307/3182614.

Haberl, H. (Ed.). (1998). *Technologische Zivilisation und Kolonisierung von Natur* (Iff-Texte, vol. 3). Wien: Springer.

Hagendorff, T. (2011). Ästhetische Wahrheitstheorien nach Rorty. *Critica—Zeitschrift für Philosophie und Kunsttheori, 2,* (2–24).

Hard, G. (2002). Über Räume reden. Zum Gebrauch des Wortes "Raum" in sozialwissenschaftlichem Zusammenhang. In G. Hard (Ed.), *Landschaft und Raum. Aufsätze zur Theorie der Geographie* (Osnabrücker Studien zur Geographie, vol. 22, pp. 235–252). Osnabrück: Universitätsverlag Rasch.

Harendt, A. (2019). *Gesellschaft. Raum. Narration. Geographische Weltbilder im Medienalltag.* Stuttgart: Franz Steiner Verlag.

Harley, J. B. (1989). Deconstructing The Map. *Cartographica: The International Journal for Geographic Information and Geovisualization, 26,* (2, 1–20). https://doi.org/10.3138/E635-7827-1757-9T53.

Hasse, J. (2007). *Übersehene Räume. Zur Kulturgeschichte und Heterotopologie des Parkhauses.* Bielefeld: transcript.

Hasse, J. (2012). *Atmosphären der Stadt. Aufgespürte Räume.* Berlin: Jovis-Verlag.

Hauser, S. (2004). Industrieareale als urbane Räume. In W. Siebel (Ed.), *Die europäische Stadt* (pp. 146–157). Frankfurt (Main): Suhrkamp.

Heiland, S. (1992). *Naturverständnis. Dimensionen des menschlichen Naturbezugs.* Darmstadt: WBG.

Heiland, S. (2010). "Kulturlandschaft". In D. Henckel, K. von Kuczkowski, P. Lau, E. Pahl-Weber, & F. Stellmacher (Eds.), *Planen—Bauen—Umwelt. Ein Handbuch* (pp. 278–283). Wiesbaden: VS Verlag für Sozialwissenschaften.

Heinrich, P. V., Miner, M., Paulsell, R., & McCulloh, R. P. (2020). *Response of Late Quaternary Valley systems to Holocene sea level rise on continental shelf offshore Louisiana: preservation potential of paleolandscapes.* Baton Rouge: U.S. Department of the Interior.

Hemmerling, S. A., Barra, M., & Bond, R. H. (2020). Adapting to a Smaller Coast: Restoration, Protection, and Social Justice in Coastal Louisiana. In S. Laska (Ed.), *Louisiana´s Response to Extreme Weather. A Coastal State's Adaptation Challenges and Successes* (pp. 113–144). Berlin, Heidelberg: Springer Nature.

Hemmerling, S. A., DeMyers, C. A., & Carruthers, T. J. B. (2022). Building Resilience through Collaborative Management of Coastal Protection and Restoration Planning in Plaquemines Parish, Louisiana, USA. *Sustainability, 14, (5).* https://doi.org/10.3390/su14052974.

Hemmerling, S. A. (2007). *Environmental equity in southeast Louisiana: oil, people, policy, and the geography of industrial hazards.* (Dissertation, Louisiana State University and Agricultural & Mechanical College). Baton Rouge. https://digitalcommons.lsu.edu/gradschool_dissertations/1304/. Accessed: 23 February 2021.

Hensel, A., Klecha, S., & Schmitz, C. (2013). "Vernetzt euch—das ist die einzige Waffe, die man hat". Internetproteste. In F. Walter, S. Marg, L. Geiges, & F. Butzlaff (Eds.), *Die neue Macht der Bürger. Was motiviert die Protestbewegungen?* (BP-Gesellschaftsstudie, pp. 267–300). Reinbek bei Hamburg: Rowohlt.

Herrmann, V. (2017). America's first climate change refugees: Victimization, distancing, and disempowerment in journalistic storytelling. *Energy Research & Social Science, 31, (2),* 205–214). https://doi.org/10.1016/j.erss.2017.05.033.

Hester, M. W., & Mendelssohn Irving A. (2000). Long-term recovery of a Louisiana brackish marsh plant community from oil-spill impact: vegetation response and mitigating effects of marsh surface elevation. *Marine Environmental Research, 49,* (*3,* 233–254).

Hester, M. W., Willis, J. M., Rouhani, S., Steinhoff, M. A., & Baker, M. C. (2016). Impacts of the Deepwater Horizon oil spill on the salt marsh vegetation of Louisiana. *Environmental Pollution,* (*216,* 361–370).

Hildebrand, D. L. (2003). The neopragmatist turn. *Southwest Philosophy Review, 19,* (*1,* 79–88).

Hildebrand, D. L. (2005). Pragmatism, neopragmatism, and public administration. *Administration & Society, 37,* (*3,* 345–359).

Hochschild, A. R. (2016). *Strangers In Their Own Land. Anger And Mourning On The American Right.* New York: The New Press.

Hoesterey, I. (2001). *Pastiche. Cultural Memory in Art, Film, Literature.* Bloomington: Indiana University Press.

Holzner, L. (1994). Geisteshaltung und Stadt-Kulturlandschaftsgestaltung. Das Beispiel der Vereinigten Staaten. *Petermanns Geographische Mitteilungen, 138,* (*1,* 51–59).

Holzner, L. (1996). *Stadtland USA. Die Kulturlandschaft des American way of life; 4 Tabellen* (Petermanns geographische Mitteilungen, Ergänzungsheft, vol. 291). Gotha: Perthes.

Houck, O. A. (2015). The Reckoning: Oil and Gas Development in the Louisiana Coastal Zone. *Tulane Environmental Law Journal,* (*28,* 185–296).

Howard, P., Thompson, I., Waterton, E., & Atha, M. (Eds.). (2019). *The Routledge Companion to Landscape Studies* (2. Auflage). London: Routledge.

Hügin, U. (1996). *Individuum, Gemeinschaft, Umwelt. Konzeption einer Theorie der Dynamik anthropogener Systeme.* Bern: Lang.

Hunziker, M. (2010). Die Bedeutung der Landschaft für den Menschen: objektive Eigenschaften der Landschaft oder individuelle Wahrnehmung des Menschen? In WSL (Ed.), *Landschaftsqualität. Konzepte, Indikatoren und Datengrundlagen* (Forum für Wissen, pp. 33–41). Birmensdorf: Eidgenössische Forschungsanstalt für Wald, Schnee und Landschaft WSL.

Hunziker, M., Felber, P., Gehring, K., Buchecker, M., Bauer, N., & Kienast, F. (2008a). Evaluation of Landscape Change by Different Social Groups. Results of Two Empirical Studies in Switzerland. *Mountain Research and Development, 28,* (*2,* 140–147). https://doi.org/10.1659/mrd.0952.

Hunziker, M., Felber, P., Gehring, K., Buchecker, M., Bauer, N., & Kienast, F. (2008b). Evaluation of Landscape Change by Different Social Groups. Results of Two Empirical Studies in Switzerland. *Mountain Research and Development, 28,* (*2,* 140–147). https://doi.org/10.1659/mrd.0952.

Imhof, E. (1972). *Thematische Kartographie.* Berlin, New York: de Gruyter.

Ipsen, D. (2006). *Ort und Landschaft.* Wiesbaden: VS Verlag für Sozialwissenschaften.

Jackson, J. B. (1984). *Discovering the Vernacular Landscape.* New Haven: Yale University Press.

James, W. (1907). *Pragmatism. An Old Way for Some New Ways of Thinking.* New York: Longmans, Green and Co.

Jekel, T., Car, A., Strobl, J., & Griesebner, G. (Eds.). (2012). *GI_Forum 2012: Geovosualization, Society and Learning. Conference Proceedings.* Berlin: Wichmann Verlag / VDE.

Jenal, C. (2019). *"Das ist kein Wald, Ihr Pappnasen!"—Zur sozialen Konstruktion von Wald. Perspektiven von Landschaftstheorie und Landschaftspraxis.* Wiesbaden: Springer VS.

Jenal, C., Kühne, O., Schäffauer, G., & Sedelmeier, T. (2021a). Louisiana und seine Herausforderungen—eine regionalgeographische Kontextualisierung. In O. Kühne, T. Sedelmeier, & C. Jenal (Eds.), *Louisiana—mediengeographische Beiträge zu einer neopragmatischen Regionalen Geographie* (pp. 43–67). Wiesbaden: Springer.

Jenal, C., Endreß, S., Kühne, O., & Zylka, C. (2021b). Technological Transformation Processes and Resistance—On the Conflict Potential of 5G Using the Example of 5G Network Expansion in Germany. *Sustainability, 13,* (*24,* 1–21). https://doi.org/10.3390/su132413550.

Jessee, N. (2020). Community Resettlement in Louisiana: Learning from Histories of Horror and Hope. In S. Laska (Ed.), *Louisiana´s Response to Extreme Weather. A Coastal State's Adaptation Challenges and Successes* (pp. 147–184). Berlin, Heidelberg: Springer Nature.

Jick, T. D. (1979). Mixing Qualitative and Quantitative Methods: Triangulation in Action. *Administrative Science Quarterly, 24,* (*4,* 602). https://doi.org/10.2307/2392366.

Jiraprasertkun, C. (2015). Thai Conceptualizations of Space, Place and Landscape. In D. Bruns, O. Kühne, A. Schönwald, & S. Theile (Eds.), *Landscape Culture—Culturing Landscapes. The Differentiated Construction of Landscapes* (pp. 95–110). Wiesbaden: Springer VS.

Jones, O. (2008). Stepping from the wreckage: Geography, pragmatism and anti-representational theory. *Geoforum, 39,* (*4,* 1600–1612). https://doi.org/10.1016/j.geoforum.2007.10.003.

Kaiser, M., & Maasen, S. (2010). Wissenschaftssoziologie. In G. Kneer, & M. Schroer (Eds.), *Handbuch Spezielle Soziologien* (pp. 685–705). Wiesbaden: Springer VS.

Kaußen, L. (2018). Landscape Perception and Construction in Social Media: An Analysis of User-generated Content. *Journal of Digital Landscape Architecture, 3,* (373–379). https://doi.org/10.14627/537642040.

Kavitha, K. M., Shetty, A., Abreo, B., D'Souza, A., & Kondana, A. (2020). Analysis and Classification of User Comments on YouTube Videos. *Procedia Computer Science, 177,* (593–598). https://doi.org/10.1016/j.procs.2020.10.084.

Kazmann, R. G., & Heath, M. M. (1986). Land Subsidence Related Ground-Water Offtake in the New Orleans Area. *Gulf Coast Association of Geological Societies Transactions, 1986,* (*8,* 108–113).

Kemp, A. C., Horton, B. P., Donnelly, J. P., Mann, M. E., Vermeer, M., & Rahmstorf, S. (2011). Climate related sea-level variations over the past two millennia. *Proceedings of the National*

Academy of Sciences of the United States of America, 108, (*27,* 11017–11022). https://doi.org/10.1073/pnas.1015619108.

Khalil, S. M. (2018). Surficial sediment distribution maps for sustainability and ecosystem restoration of coastal Louisiana. *Shore and Beach, 86,* (*3,* 21–29).

Kiefl, W. (2001). Erlebnis Strandurlaub. Ergebnisse einer Beobachtungsstudie. In A. G. Keul, R. Bachleitner, & H. J. Kagelmann (Eds.), *Gesund durch Erleben? Beiträge zur Erforschung der Tourismusgesellschaft* (2nd ed., pp. 132–138). München, Wien: Profil.

Kim, A. M. (2015). Critical cartography 2.0: From "participatory mapping" to authored visualizations of power and people. *Landscape and Urban Planning, 142,* (215–225). https://doi.org/10.1016/j.landurbplan.2015.07.012.

King, M. (2017). A tribe faces rising tides: the resettlement of Isle de Jean Charles. *LSU Journal of Energy Law and Resources, 6,* (*1,* 295–317).

Kitchin, R., & Dodge, M. (2011). *Code/Space. Software and Everyday Life.* Cambridge MA: The MIT Press.

Kittler, T. (2021). Mediendiskurs zum Klimawandel in Louisiana—eine quantitative Analyse von YouTube-Kommentaren. In O. Kühne, T. Sedelmeier, & C. Jenal (Eds.), *Louisiana—mediengeographische Beiträge zu einer neopragmatischen Regionalen Geographie* (pp. 167–181). Wiesbaden: Springer.

Kneer, G., & Nassehi, A. (1997). *Niklas Luhmanns Theorie sozialer Systeme. Eine Einführung.* München: Fink.

Knoblauch, H. (2003). Habitus und Habitualisierung. Zur Komplementarität von Bourdieu mit dem Sozialkonstruktivismus. In B. Rehbein, G. Saalmann, & H. Schwengel (Eds.), *Pierre Bourdieus Theorie des Sozialen. Probleme und Perspektiven* (pp. 187–201). Konstanz: UVK Verlagsgesellschaft.

Koegst, L. (2022). Über drei Welten, Räume und Landschaften. Digital geführte Exkursionen an Hochschulen aus der Perspektive der drei Welten Theorie im Allgemeinen und der Theorie der drei Landschaften im Speziellen. *Berichte Geographie und Landeskunde, 69,* (*3,* 1–21). https://doi.org/10.25162/bgl-2022-0012.

Kolb, C. R., & Saucier, R. T. (1982). Engineering Geology of New Orleans. *Reviews in Engineering Geology,* (*5,* 75–93).

Kolker, A. S., Alison, M. A., & Hameed, S. (2011). An evaluation of subsidence rates and sea-level variability in the northern Gulf of Mexico. *Geophysical Research Letters, 38, (21).* https://doi.org/10.1029/2011GL049458.

Korf, B. (2021). 'German Theory': On Cosmopolitan geographies, counterfactual intellectual histories and the (non)travel of a 'German Foucault'. *Environment and Planning D: Society and Space),* (026377582198969). https://doi.org/10.1177/0263775821989697.

Kostner, S. (2019). Identitätslinke Läuterungsagenda. Welche Folgen hat sie für Migrationsgesellschaften? In S. Kostner (Ed.), *Identitätslinke Läuterungsagenda. Eine Debatte zu ihren Folgen für Migrationsgesellschaften* (17–74). Stuttgart: Ibidem-Verlag.

Kovaka, K. (2021). Climate change denial and beliefs about science. *Synthese, 198,* (*3,* 2355–2374). https://doi.org/10.1007/s11229-019-02210-z.

Kuckartz, U. (2014). *Mixed Methods. Methodologie, Forschungsdesigns und Analyseverfahren.* Wiesbaden: Springer VS.

Kühne, O. (2006a). *Landschaft in der Postmoderne. Das Beispiel des Saarlandes.* Wiesbaden: DUV.

Kühne, O. (2006b). Soziale Distinktion und Landschaft. Eine landschaftssoziologische Betrachtung. *Stadt+Grün,* (*12,* 42–45).

Kühne, O. (2008). *Distinktion—Macht—Landschaft. Zur sozialen Definition von Landschaft.* Wiesbaden: VS Verlag für Sozialwissenschaften.

Kühne, O. (2012a). *Stadt—Landschaft—Hybridität. Ästhetische Bezüge im postmodernen Los Angeles mit seinen modernen Persistenzen.* Wiesbaden: Springer VS.
Kühne, O. (2012b). Urban nature between modern and postmodern aesthetics: Reflections based on the social constructivist approach. *Quaestiones Geographicae, 31,* (*2,* 61–70). https://doi.org/10.2478/v10117-012-0019-3.
Kühne, O. (2018a). Die Landschaften 1, 2 und 3 und ihr Wandel. Perspektiven für die Landschaftsforschung in der Geographie—50 Jahre nach Kiel. *Berichte. Geographie und Landeskunde, 92,* (*3–4,* 217 – 231).
Kühne, O. (2018b). Die Moralisierung von Landschaft—Überlegungen zu einer problematischen Kommunikation aus Sicht der Luhmannschen Systemtheorie. In S. Hennecke, H. Kegler, K. Klaczynski, & D. Münderlein (Eds.), *Diedrich Bruns wird gelehrt haben. Eine Festschrift* (pp. 115–121). Kassel: Kassel University Press.
Kühne, O. (2018c). *Landscape and Power in Geographical Space as a Social-Aesthetic Construct.* Dordrecht: Springer International Publishing.
Kühne, O. (2018d). *Landschaft und Wandel. Zur Veränderlichkeit von Wahrnehmungen.* Wiesbaden: Springer VS.
Kühne, O. (2018e). Reboot "Regionale Geographie"—Ansätze einer neopragmatischen Rekonfiguration "horizontaler Geographien". *Berichte. Geographie und Landeskunde, 92,* (*2,* 101–121).
Kühne, O. (2019a). Die Produktivität von Landschaftskonflikten—Möglichkeiten und Grenzen auf Grundlage der Konflikttheorie Ralf Dahrendorfs. In K. Berr, & C. Jenal (Eds.), *Landschaftskonflikte* (pp. 37–49). Wiesbaden: Springer VS.
Kühne, O. (2019b). *Landscape Theories. A Brief Introduction.* Wiesbaden: Springer VS.
Kühne, O. (2020a). Landscape Conflicts. A Theoretical Approach Based on the Three Worlds Theory of Karl Popper and the Conflict Theory of Ralf Dahrendorf, Illustrated by the Example of the Energy System Transformation in Germany. *Sustainability: Science, Practice and Policy, 12,* (*17,* 1–20). https://doi.org/10.3390/su12176772.
Kühne, O. (2020b). The Social Construction of Space and Landscape in Internet Videos. In D. Edler, C. Jenal, & O. Kühne (Eds.), *Modern Approaches to the Visualization of Landscapes* (pp. 121–137). Wiesbaden: Springer VS.
Kühne, O. (2021a). Contours of a 'Post-Critical' Cartography—A Contribution to the Dissemination of Sociological Cartographic Research. *KN—Journal of Cartography and Geographic Information,* (1–9). https://doi.org/10.1007/s42489-021-00080-5.
Kühne, O. (2021b). *Landschaftstheorie und Landschaftspraxis. Eine Einführung aus sozialkonstruktivistischer Perspektive* (3., aktualisierte und überarbeitete Auflage). Wiesbaden: Springer VS.
Kühne, O. (2021c). Louisiana—mediengeographische Beiträge zu einer neopragmatischen Regionalen Geographie. Eine Einführung. In O. Kühne, T. Sedelmeier, & C. Jenal (Eds.), *Louisiana—mediengeographische Beiträge zu einer neopragmatischen Regionalen Geographie* (pp. 1–11). Wiesbaden: Springer.
Kühne, O. (2021d). Potentials of the Three Spaces Theory for Understandings of Cartography, Virtual Realities, and Augmented Spaces. *KN—Journal of Cartography and Geographic Information), 71,* (*4,* 297–305). https://doi.org/10.1007/s42489-021-00089-w.
Kühne, O. (2022a). ‚Neopragmatische' thematische Kartographie—Grundzüge ihrer Theorie, Begründung und Anwendung. In O. Kühne, C. Jenal, & T. Sedelmeier (Eds.), *Cultural Atlas of TÜbingenness. Kleine Karten aus dem großen TÜbiversum* (pp. 11–21). Wiesbaden: Springer Fachmedien Wiesbaden GmbH; Springer VS.
Kühne, O. (2022b). Representations of landscape in the strategy game Civilization. In D. Edler, O. Kühne, & C. Jenal (Eds.), *The Social Construction of Landscape in Games* (261–272). Wiesbaden: Springer.

Kühne, O. (2022c). Foodscapes – a Neopragmatic Redescription. Berichte. *Geographie und Landeskunde*, (online first, 1–21). https://doi.org/10.25162/bgl-2022-0016

Kühne, O., & Berr, K. (2021). *Wissenschaft, Raum, Gesellschaft. Eine Einführung zur sozialen Erzeugung von Wissen.* Wiesbaden: Springer VS.

Kühne, O., & Edler, D. (2018). Multisensorische Landschaften—die Bedeutung des Nicht-Visuellen bei der sozialen und individuellen Konstruktion von Landschaft und Herausforderungen für ihre Erfassung und Wiedergabe. *Berichte. Geographie und Landeskunde, 92,* (*1,* 27–45).

Kühne, O., & Edler, D. (2022). Georg Simmel Goes Virtual. From 'Philosophy of Landscape' to the Possibilities of Virtual Reality in Landscape Research. *Societies, 12,* (*5,* 122). https://doi.org/10.3390/soc12050122.

Kühne, O., & Jenal, C. (2020a). *Baton Rouge—The Multivillage Metropolis. A Neopragmatic Landscape Biographical Approach on Spatial Pastiches, Hybridization, and Differentiation.* Wiesbaden: Springer VS.

Kühne, O., & Jenal, C. (2020b). Baton Rouge (Louisiana): On the Importance of Thematic Cartography for 'Neopragmatic Horizontal Geography'. *KN—Journal of Cartography and Geographic Information, 71,* (*1,* 23–31). https://doi.org/10.1007/s42489-020-00054-z.

Kühne, O., & Jenal, C. (2020c). The Threefold ´Landscape Dynamics—Basic Considerations, Conflicts and Potentials of Virtual Landscape Research. In D. Edler, C. Jenal, & O. Kühne (Eds.), *Modern Approaches to the Visualization of Landscapes* (pp. 389–402). Wiesbaden: Springer VS.

Kühne, O., & Jenal, C. (2021a). Baton Rouge—A Neopragmatic Regional Geographic Approach. *Urban Science, 5,* (*1,* 1–17). https://doi.org/10.3390/urbansci5010017.

Kühne, O., & Jenal, C. (2021b). Neopragmatische Regionale Geographien—eine Annäherung. In O. Kühne, T. Sedelmeier, & C. Jenal (Eds.), *Louisiana—mediengeographische Beiträge zu einer neopragmatischen Regionalen Geographie* (pp. 13–23). Wiesbaden: Springer.

Kühne, O., & Koegst, L. (2022). Cartographic Representations of Coastal Land Loss in Louisiana. An Investigation Based on Deviant Cartographies. *KN—Journal of Cartography and Geographic Information, 2022,* (*4,* 1–12). https://doi.org/10.1007/s42489-022-00120-8.

Kühne, O., & Schönwald, A. (2015). *San Diego. Eigenlogiken, Widersprüche und Hybriditäten in und von ‚America's finest city'.* Wiesbaden: Springer VS.

Kühne, O., & Weber, F. (2015). Der Energienetzausbau in Internetvideos—eine quantitativ ausgerichtete diskurstheoretisch orientierte Analyse. In S. Kost, & A. Schönwald (Eds.), *Landschaftswandel—Wandel von Machtstrukturen* (pp. 113–126). Wiesbaden: Springer VS.

Kühne, O., & Weber, F. (2019). *Hybrid California. Annäherungen an den Golden State, seine Entwicklungen, Ästhetisierungen und Inszenierungen.* Wiesbaden: Springer VS.

Kühne, O., Weber, F., & Berr, K. (2019). The productive potential and limits of landscape conflicts in light of Ralf Dahrendorf's conflict theory. *Società Mutamento Politica, 10,* (*19,* 77–90). https://oajournals.fupress.net/index.php/smp/article/view/10597. Accessed: 22 June 2020.

Kühne, O., Jenal, C., & Edler, D. (2020). Functions of Landscape in Games—A Theoretical Approach with Case Examples. *Arts, 9 (4).* https://doi.org/10.3390/arts9040123.

Kühne, O., Jenal, C., & Koegst, L. (2020). Postmoderne Siedlungsentwicklungen in Baton Rouge, Louisiana: Stadtlandhybridität und Raumpastiches zwischen Begrenzungen und Entgrenzungen. In F. Weber, C. Wille, B. Caesar, & J. Hollstegge (Eds.), *Geographien der Grenzen. Räume—Ordnungen—Verflechtungen* (391–411). Wiesbaden: Springer VS.

Kühne, O., Koegst, L., Zimmer, M.-L., & Schäffauer, G. (2021). "… Inconceivable, Unrealistic and Inhumane". Internet Communication on the Flood Disaster in West Germany of July 2021 between Conspiracy Theories and Moralization—A Neopragmatic Explorative Study. *Sustainability, 13,* (*20,* 1–23). https://doi.org/10.3390/su132011427.

Kühne, O., Edler, D., & Jenal, C. (2021a). A Multi-Perspective View on Immersive Virtual Environments (IVEs). *ISPRS—International Journal of Geo-Information, 10,* (*8,* 1–22). https://doi.org/10.3390/ijgi10080518.

Kühne, O., Berr, K., Schuster, K., & Jenal, C. (2021). *Freiheit und Landschaft. Auf der Suche nach Lebenschancen mit Ralf Dahrendorf.* Wiesbaden: Springer.

Kühne, O., Berr, K., Jenal, C., & Schuster, K. (2021). *Liberty and Landscape. In Search of Life-chances with Ralf Dahrendorf.* Basingstoke: Palgrave Macmillan.

Kühne, O., Jenal, C., Schäffauer, G., & Sedelmeier, T. (2021). Louisianas Weg zum Bundesstaat der ‚multiplen Herausforderungen'—eine historische Kontextualisierung. In O. Kühne, T. Sedelmeier, & C. Jenal (Eds.), *Louisiana—mediengeographische Beiträge zu einer neopragmatischen Regionalen Geographie* (pp. 25–41). Wiesbaden: Springer.

Kühne, O., Edler, D., & Jenal, C. (2021b). The Abstraction of an Idealization: Cartographic Representations of Model Railroads. Die Abstraktion der Idealisierung—über kartographische Repräsentationen von Modellbahnlandschaften. *KN—Journal of Cartography and Geographic Information, 71,* (*2,* 207–217). https://doi.org/10.1007/s42489-020-00064-x.

Kühne, O., Schönwald, A., & Jenal, C. (2022). Bottom-up memorial landscapes between social protest and top-down tourist destination: the case of Chicano Park in San Diego (California)—an analysis based on Ralf Dahrendorf's conflict theory. *Landscape Research,* (1–17). https://doi.org/10.1080/01426397.2022.2069731.

Kühne, O., Parush, D., Shmueli, D., & Jenal, C. (2022). Conflicted Energy Transition—Conception of a Theoretical Framework for Its Investigation. *Land, 11,* (*1,* 116). https://doi.org/10.3390/land11010116.

Kühne, O., Berr, K., & Jenal, C. (2022). *Die geschlossene Gesellschaft und ihre Ligaturen. Eine Kritik am Beispiel 'Landschaft'.* Wiesbaden: Springer VS.

Kühne, O., Jenal, C., & Edler, D. (2022). Landscapes in Games. Insights and Overviews on Contingencies between Worlds 1, 2 and 3. In D. Edler, O. Kühne, & C. Jenal (Eds.), *The Social Construction of Landscape in Games* (pp. 77–87). Wiesbaden: Springer.

Kühne, O., Edler, D., & Jenal, C. (2022a). Landschaften und Spiele—von Virtualisierungen, Hybridisierungen und der Steigerung von Kontingenz. *Berichte. Geographie und Landeskunde, 96,* (in Veröfentlichung).

Kühne, O., Jenal, C., & Berger, S. (2022). Saarbrücken-Brebach—phänomenologische Zugänge zu einem touristisch kaum präsenten, altindustriell geprägten Ort. In O. Kühne, T. Freytag, T. Sedelmeier, & C. Jenal (Eds.), *Landschaft und Tourismus* (RaumFragen, in diesem Band). Wiesbaden: Springer.

Kühne, O., Denzer, V., & Eissner, C. (2022). The beach in the box—aspects of the construction and experience of a hybrid landscape. In D. Edler, O. Kühne, & C. Jenal (Eds.), *The Social Construction of Landscape in Games* (pp. 163–179). Wiesbaden: Springer.

Kühne, O., Edler, D., & Jenal, C. (2022b). The cartographic representation of model railroad landscapes—theoretical considerations and empirical results from model railroad-related literature. In D. Edler, O. Kühne, & C. Jenal (Eds.), *The Social Construction of Landscape in Games* (pp. 127–148). Wiesbaden: Springer.

Lamb, Z. (2020). Connecting the dots: The origins, evolutions, and implications of the map that changed post-Katrina recovery planning in New Orleans. In S. Laska (Ed.), *Louisiana's Response to Extreme Weather. A Coastal State's Adaptation Challenges and Successes* (Springer eBooks Earth and Environmental Science, 1st ed., pp. 65–91). Cham: Springer.

Laska, S., Peterson, K., Rodrigue, C., Cosse, T., Philippe, R., Burchett, O., et al. (2015). Layering of natural and human caused disasters in the context of anticipated climate change disasters: The coastal Louisiana experience. In M. Companion (Ed.), *Disaster's impact on livelihood and cultural survival. Losses, opportunities, and mitigation* (pp. 225–238). Boca Raton, Florida: CRC Press.

Latour, B. (1993). *We Have Never Been Modern.* New York: Harvester Wheatsheaf.

Laudan, L. (1977). *Progress and Its Problems. Towards a Theory of Scientific Growth.* Berkeley, Calif: University of California Press.

Lefebvre, M. (Ed.). (2006). *Landscape and Film.* New York: Routledge.

Lefkowitz Horowitz, H. (2015). The Writer's Path. JB Jackson and Cultural Geography as a Literary Genre. *SiteLINES: A Journal of Place, 11,* (*1,* 3–7).

Leibenath, M. (2014). Landschaftsbewertung im Spannungsfeld von Expertenwissen, Politik und Macht. *UVP-report, 28,* (*2,* 44–49). https://www2.ioer.de/recherche/pdf/2014_leibenath_uvp-report.pdf. Accessed: 26 January 2017.

Lemos, A. (2010). Post—Mass Media Functions, Locative Media, and Informational Territories: New Ways of Thinking About Territory, Place, and Mobility in Contemporary Society. *Space and Culture, 13,* (*4,* 403–420). https://doi.org/10.1177/1206331210374144.

Lenček, L., & Bosker, G. (1998). *The beach. The history of paradise on earth.* London u. a.: Secker & Warburg.

Linke, S. (2017). Ästhetik, Werte und Landschaft—eine Betrachtung zwischen philosophischen Grundlagen und aktueller Praxis der Landschaftsforschung. In O. Kühne, H. Megerle, & F. Weber (Eds.), *Landschaftsästhetik und Landschaftswandel* (RaumFragen: Stadt—Region—Landschaft, pp. 23–40). Wiesbaden: Springer VS.

Linke, S. (2020). Landscape in Internet Pictures. In D. Edler, C. Jenal, & O. Kühne (Eds.), *Modern Approaches to the Visualization of Landscapes* (pp. 139–156). Wiesbaden: Springer VS.

Linke, S. I. (2019). *Die Ästhetik medialer Landschaftskonstrukte. Theoretische Reflexionen und empirische Befunde.* Wiesbaden: Springer VS.

Ljunge, M. (2013). Beyond 'the Phenomenological Walk': Perspectives on the Experience of Images. *Norwegian Archaeological Review, 46,* (*2,* 139–158). https://doi.org/10.1080/00293652.2013.821160.

Loda, M., Kühne, O., & Puttilli, M. (2020). The Social Construction of Tuscany in the German and English Speaking World—Presented by the Analysis of Internet Images. In D. Edler, C. Jenal, & O. Kühne (Eds.), *Modern Approaches to the Visualization of Landscapes* (pp. 157–171). Wiesbaden: Springer VS.

Löffler, E. (1985). *Geographie und Fernerkundung. Eine Einführung in die Geographische Interpretation von Luftbildern und modernen Fernerkundungsdaten.* Wiesbaden: Springer Vieweg.

Löfgren, O. (2002). *On Holiday. A History of Vacationing.* Berkeley: University of California Press.

Luhmann, N. (1986). *Ökologische Kommunikation. Kann die moderne Gesellschaft sich auf ökologische Gefährdungen einstellen?* Opladen: Westdeutscher Verlag.

Luhmann, N. (1987). *Soziale Systeme. Grundriß einer allgemeinen Theorie* (Suhrkamp-Taschenbuch Wissenschaft, vol. 666). Frankfurt am Main: Suhrkamp.

Luhmann, N. (1990). *Die Wissenschaft der Gesellschaft.* Frankfurt a. M.: Suhrkamp.

Luhmann, N. (1993). Die Moral des Risikos und das Risiko der Moral. In G. Bechmann (Ed.), *Risiko und Gesellschaft. Grundlagen und Ergebnisse interdisziplinärer Risikoforschung* (pp. 327–338). Opladen: Westdeutscher Verlag.

Luhmann, N. (1996). *Die Realität der Massenmedien.* Opladen: Westdeutscher Verlag.

Luhmann, N. (2016). *Die Moral der Gesellschaft* (Suhrkamp-Taschenbuch Wissenschaft, vol. 1871, 4. Aufl.). Frankfurt am Main: Suhrkamp.

Lukinbeal, C. (2012). "On Location" Filming in San Diego County from 1985–2005: How a Cinematic Landscape Is Formed Through Incorporative Tasks and Represented Through Mapped

Inscriptions. *Annals of the Association of American Geographers, 102,* (*1,* 171–190). https://doi.org/10.1080/00045608.2011.583574.

Mackert, J. (2010). Opportunitätsstrukturen und Lebenschancen. *Berliner Journal für Soziologie, 20,* (*3,* 401–420). https://doi.org/10.1007/s11609-010-0135-7.

Macpherson, H. (2016). Walking methods in landscape research: moving bodies, spaces of disclosure and rapport. *Landscape Research, 41,* (*1,* 425–432). https://doi.org/10.1080/01426397.2016.1156065.

Madden, A., Ruthven, I., & McMenemy, D. (2013). A classification scheme for content analyses of YouTube video comments. *Journal of Documentation, 69,* (*5,* 693–714). https://doi.org/10.1108/JD-06-2012-0078.

Makhzoumi, J. M. (2002). Landscape in the Middle East: An inquiry. *Landscape Research, 27,* (*3,* 213–228). https://doi.org/10.1080/01426390220149494.

Maldonado, J. K. (2019). *Seeking Justice in an Energy Sacrifice Zone. Standing on Vanishing Land in Coastal Louisiana.* Abingdon: Routledge.

Maldonado, J. K. (2015). Everyday practices and symbolic forms of resistance: adapting to environmental change in coastal Louisiana. In A. E. Collins (Ed.), *Hazards, risks, and disasters in society* (Hazards and Disasters Series, pp. 199–216). Amsterdam, Netherlands: Elsevier.

Mann, M., & Rauscher, S. (2021). Schulsegregation in Louisiana—die Behandlung von Schulsegregation auf sozialen Videoplattformen am Beispiel des Videos "How Black High School Students Are Hurt by Modern-Day Segregation". In O. Kühne, T. Sedelmeier, & C. Jenal (Eds.), *Louisiana—mediengeographische Beiträge zu einer neopragmatischen Regionalen Geographie* (pp. 133–150). Wiesbaden: Springer.

Massey, D. B. (2013). *Space, Place and Gender.* New York: John Wiley & Sons.

Mattissek, A., & Wiertz, T. (2014). Materialität und Macht im Spiegel der Assemblage-Theorie: Erkundungen am Beispiel der Waldpolitik in Thailand. *Geographica Helvetica, 69,* (*3,* 157–169).

Mayring, P. (2008). *Qualitative Inhaltsanalyse. Grundlagen und Techniken* (10., neu ausgestattete Auflage). Weinheim: Beltz.

McBride, R. A., Taylor, M. J., & Byrnes, M. R. (2007). Coastal morphodynamics and Chenier-Plain evolution in southwestern Louisiana, USA: A geomorphic model. *Geomorphology, 88,* (*3–4,* 367–422). https://doi.org/10.1016/j.geomorph.2006.11.013.

Medyńska-Gulij, B. (2012). Pragmatische Kartographie in Google Maps API. *KN—Journal of Cartography and Geographic Information, 62,* (*5,* 250–255). https://doi.org/10.1007/BF03544493.

Menand, L. (2001). *The Metaphysical Club. A story of ideas in America* (1.th ed.). New York, NY: Farrar Straus and Giroux.

Miggelbrink, J. (2014). Diskurs, Machttechnik, Assemblage. Neue Impulse für eine regionalgeographische Forschung. *Geographische Zeitschrift, 102,* (*1,* 25–40).

Molnar, P. H. (2015). *Plate tectonics. A very short introduction* (Very short introductions, vol. 425). Oxford: Oxford University Press.

Moore, N. R. (1972). *Improvement of the Lower Mississippi River and tributaries, 1931–1972* (1st ed.). Vicksburg, Mississippi: Mississippi River Commission.

Morse, J. M. (1991). Approaches to Qualitative-Quantitative Methodological Triangulation. *Nursing Research, 40,* (*2,* 120–123). https://doi.org/10.1097/00006199-199103000-00014.

Morton, R. A., Bernier, J. C., & Barras, J. A. (2006). Evidence of regional subsidence and associated interior wetland loss induced by hydrocarbon production, Gulf Coast region, USA. *Environmental Geology, 50,* (*2,* 261–274). https://doi.org/10.1007/s00254-006-0207-3.

Mossa, J. (1996). Sediment dynamics in the lowermost Mississippi River. *Engineering Geology, 45,* (*1–4,* 457–479). https://doi.org/10.1016/S0013-7952(96)00026-9.

Mounce, H. O. (2002). *The two pragmatisms. From Peirce to Rorty.* London: Routledge.

Mousavi, M. E., Irish, J. L., Frey, A. E., Olivera, F., & Edge, B. L. (2011). Global warming and hurricanes: the potential impact of hurricane intensification and sea level rise on coastal flooding. *Climatic Change, 104,* (*3–4,* 575–597). https://doi.org/10.1007/s10584-009-9790-0.

Müller, K. (2013). *Allgemeine Systemtheorie. Geschichte, Methodologie und sozialwissenschaftliche Heuristik eines Wissenschaftsprogramms.* Wiesbaden: VS Verlag für Sozialwissenschaften.

Müller, M. (2015). Assemblages and Actor-networks: Rethinking Socio-material Power, Politics and Space. *Geography Compass, 9,* (*1,* 27–41). https://doi.org/10.1111/gec3.12192.

Müller, M. (2021). *Rorty lesen.* Wiesbaden: Springer VS.

Nagle, A. (2017). *Kill all normies. The online culture wars from Tumblr and 4chan to the alt-right and Trump.* Winchester, UK: Zero Books.

NASA. (2022). Worldview, Louisiana, Satellite Image 2022 May 12. https://worldview.earthdata.nasa.gov/?v=-92.074812890625,28.5876564453125,-88.699812890625,30.1499123046875&l=Reference_Labels_15m,Reference_Features_15m,Coastlines_15m,VIIRS_NOAA20_CorrectedReflectance_TrueColor,VIIRS_SNPP_CorrectedReflectance_TrueColor,MODIS_Aqua_CorrectedReflectance_TrueColor,MODIS_Terra_CorrectedReflectance_TrueColor&lg=true&t=2022-05-11-T12%3A00%3A00Z.

Nations Online. (2022). Map of Louisiana (LA). https://www.nationsonline.org/oneworld/map/USA/louisiana_map.htm. Accessed: 14 May 2022.

Neubert, S. (2004). Pragmatismus—thematische Vielfalt in Deweys Philosophie und in ihrer heutigen Rezeption. In L. A. Hickman, S. Neubert, & K. Reich (Eds.), *John Dewey. Zwischen Pragmatismus und Konstruktivismus* (Interaktionistischer Konstruktivismus, Band 1, pp. 13–27). Münster: Waxmann.

Neverla, I., Taddicken, M., Lörcher, I., & Hoppe, I. (Eds.). (2019). *Klimawandel im Kopf. Studien zur Wirkung, Aneignung und Online-Kommunikation.* Wiesbaden: Springer Fachmedien.

Nida-Rümelin, J. (2020). *Die gefährdete Rationalität der Demokratie. Ein politischer Traktat.* Hamburg: Edition Körber.

Niedenzu, H.-J. (2001). Konflikttheorie: Ralf Dahrendorf. In J. Morel, E. Bauer, T. Maleghy, H.-J. Niedenzu, M. Preglau, & H. Staubmann (Eds.), *Soziologische Theorie. Abriß ihrer Hauptvertreter* (7th ed., pp. 171–189). München: R. Oldenbourg Verlag.

Niemann, H.-J. (2019). Karl Poppers Spätwerk und seine ‚Welt 3‘. In G. Franco (Ed.), *Handbuch Karl Popper* (pp. 1–18). Wiesbaden: Springer Reference Geisteswissenschaften.

Nittrouer, J. A., Best, J. L., Brantley, C., Cash, R. W., Czapiga, M., Kumar, P., et al. (2012). Mitigating land loss in coastal Louisiana by controlled diversion of Mississippi River sand. *Nature Geoscience, 5,* (*8,* 534–537). https://doi.org/10.1038/ngeo1525.

Olea, R. A., & Coleman, J. L. (2014). A Synoptic Examination of Causes of Land Loss in Southern Louisiana as Related to the Exploitation of Subsurface Geologic Resources. *Journal of Coastal Research, 297,* (1025–1044). https://doi.org/10.2112/JCOASTRES-D-13-00046.1.

Osbaldiston, N. (2018). *Towards a sociology of the coast. Our past, present and future relationship to the shore.* London: Palgrave Macmillan.

Papadimitriou, F. (2020). Modelling and Visualization of Landscape Complexity with Braid Topology. In D. Edler, C. Jenal, & O. Kühne (Eds.), *Modern Approaches to the Visualization of Landscapes* (pp. 79–101). Wiesbaden: Springer VS.

Papadimitriou, F. (2021). *Spatial Complexity. Theory, mathematical methods and applications.* Cham: Springer Nature.

Papadimitriou, F. (2022). *Spatial Entropy and Landscape Analysis.* Wiesbaden: Springer Fachmedien Wiesbaden.

Paris, R. (2005). *Normale Macht. Soziologische Essays.* Konstanz: UVK Verlagsgesellschaft.

Peet, R. (1998). *Modern Geographic Thought.* Oxford: Blackwell Publishers.

Peterson, K., & Maldonado, J. K. (2006). When Adaptation Is Not Enough: Between the "Now and Then" of Community-Led Resettlement. In S. Crate, & M. Nuttall (Eds.), *Anthropology and climate change. From actions to transformations* (336–353). New York: Routledge Taylor & Francis Group.

Peterson, K. J. (2020). Sojourners in a New Land: Hope and Adaptive Traditions. In S. Laska (Ed.), *Louisiana´s Response to Extreme Weather. A Coastal State's Adaptation Challenges and Successes* (pp. 185–214). Berlin, Heidelberg: Springer Nature.

Poerting, J., & Marquardt, N. (2019). Kritisch-geographische Perspektiven auf Landschaft. In O. Kühne, F. Weber, K. Berr, & C. Jenal (Eds.), *Handbuch Landschaft* (pp. 145–152). Wiesbaden: Springer VS.

Popper, K. R. (1963). *Conjectures and refutations. The growth of scientific knowledge*. London: Routledge & Kegan.

Popper, K. R. (1979). Three Worlds. Tanner Lecture, Michigan, April 7, 1978. *Michigan Quarterly Review, (1,* 141–167). https://tannerlectures.utah.edu/_documents/a-to-z/p/popper80.pdf. Accessed: 12 May 2020.

Popper, K. R. (1984). *Auf der Suche nach einer besseren Welt. Vorträge und Aufsätze aus dreißig Jahren*. München: Piper.

Popper, K. R. (1992). *Die offene Gesellschaft und ihre Feinde. Falsche Propheten—Hegel Marx und die Folgen* (vol. 2, 7th ed., 2 vols.). Tübingen: J. C. B. Mohr (Original work published 1945).

Popper, K. R. (1996). *Alles Leben ist Problemlösen. Über Erkenntnis, Geschichte und Politik*. München: Piper.

Popper, K. R. (2011[1947]). *The Open Society and Its Enemies*. Abingdon: Routledge.

Popper, K. R., & Eccles, J. C. (1977). *Das Ich und sein Gehirn*. München: Piper.

Popper, K. R., Lorenz, K., Kreuzer, F., & Sexl, R. (1994). *Die Zukunft ist offen. Das Altenberger Gespräch. Mit den Texten des Wiener Popper-Symposiums* (6. Auflage). München: Piper.

Preston-Whyte, R. (2004). The beach as a liminal space. In A. A. Lew, A. M. Williams, & C. M. Hall (Eds.), *A companion to tourism* (Blackwell companions to geography, pp. 349–359). Malden, Mass.: Blackwell Publishers.

Priest, T., & Theriot, J. (2013). Who Destroyed the Marsh?: Oil Field Canals, Coastal Ecology, and the Debate over Louisiana's Shrinking Wetlands. In J. Allured, & M. S. Martin (Eds.), *Louisiana Legacies. Readings in the History of the Pelican State* (pp. 331–341). Chichester: Wiley-Blackwell.

Purcell, D. (2018). The Internet. In P. C. Adams, J. Craine, & J. Dittmer (Eds.), *The Ashgate Research Companion to Media Geography* (pp. 137–152). London: Routledge.

Putnam, H. (1995). *Pragmatism: An Open Question*. Oxford: Blackwell Publishers.

Raab, J. (2001). *Soziologie des Geruchs. Über die soziale Konstruktion olfaktorischer Wahrnehmung*. Konstanz: UVK Verlagsgesellschaft.

Rand McNally. (2019). *Louisiana State Map*. Chicago, Illinois: Rand McNally.

Reese-Schäfer, W. (2016). *Richard Rorty zur Einführung*. Hamburg: Junius Verlag.

Riesman, D. (1950). *The Lonely Crowd*. New Haven: Yale University Press.

Rodaway, P. (1994). *Sensuous Geographies. Body, Sense, and Place*. London: Routledge.

Rodaway, P. (2002). *Sensuous Geographies: Body, Sense, and Place*. London: Routledge.

Rogers, J. D., Boutwell, G., Watkins, C., & Karadeniz, D. (2006). Geology of the New Orleans Region. In Seed, R. B. et al. (Ed.), *Investigation of the Performance of the New Orleans Flood Protection Systems in Hurricane Katrina on August 29, 2005. Final Report* (vol. 1, III1–III49). New Orleans: Self-Publishing.

Ropohl, G. (2012). *Allgemeine Systemtheorie. Einführung in transdisziplinäres Denken*. Berlin: Edition Sigma.

Rorty, R. (1981). *Der Spiegel der Natur: Eine Kritik der Philosophie*. Frankfurt am Main: Suhrkamp.

Rorty, R. (1982). *Consequences of Pragmatism. Essays: 1972–1980*. Minneapolis: University of Minnesota Press.

Rorty, R. (1989). *Kontingenz, Ironie und Solidarität*. Frankfurt (Main): Suhrkamp.

Rorty, R. (1991). *Objectivity, Relativism, and Truth*. Cambridge: Cambridge University Press.

Rorty, R. (1994). *Hoffnung statt Erkenntnis. Eine Einführung in die pragmatische Philosophie*. Wien: Passagen-Verlag.

Rorty, R. (1997). *Contingency, irony, and solidarity* (Reprint). Cambridge: Cambridge University Press.

Rorty, R. (1998). The Contingency of Language. In M. F. Bernard-Donals, & R. R. Glejzer (Eds.), *Rhetoric in an antifoundational World. Language, culture, and pedagogy* (pp. 65–85). New Haven, Connecticut: Yale University Press.

Rorty, R. (2001a). Erwiderung auf Friederike Müller-Frimauth. In T. Tietz, & U. Schäfer (Eds.), *Hinter den Spiegeln. Beiträge zur Philosophie Richard Rortys mit Erwiderungen von Richard Rorty* (pp. 259–263). Frankfurt am Main: Suhrkamp.

Rorty, R. (2001b). Erwiderung auf Hauke Brunkhorst. In T. Tietz, & U. Schäfer (Eds.), *Hinter den Spiegeln. Beiträge zur Philosophie Richard Rortys mit Erwiderungen von Richard Rorty* (pp. 162–165). Frankfurt am Main: Suhrkamp.

Rüthers, B. (1999). Einführung. Medien als vierte Gewalt. In G. von Graevenitz, R. Köcher, & B. Rüthers (Eds.), *Vierte Gewalt? Medien und Medienkontrolle. 16. Baden-Württemberg-Kolloquium* (pp. 11–18). Konstanz: UVK-Medien.

Saucier, R. T. (1994). *Geomorphology and Quaternary Geologic History of the Lower Mississippi Valley*. Vicksburg, Mississippi: U.S. Army Engineer Waterways Experiment Station.

Sayer, A. (2010). *Method in Social Science. A Realist Approach*. London, New York: Routledge.

Scaife, W. W., Turner, R. E., & Costanza, R. (1983). Coastal Louisiana recent land loss and canal impacts. *Environmental Management, 7,* (*5,* 433–442). https://doi.org/10.1007/BF01867123.

Schafranek, M., Huber, F., & Werndl, C. (2006). Die evolutionäre Grundlage Poppers Drei-Welten-Lehre. Eine unberücksichtigte Perspektive in der human-ökologischen Theoriendiskussion der Geographie, *94,* (*3,* 129–142). https://www.jstor.org/stable/27819084?casa_token=bsa1rjnquh0aaaaa:cxgw0_vhrumejewsom8umrj2ggy-dqnvwreiydbkv5wuysf1unw3nmdl5iglrhqu7-1vj-buazkebayfee7vygxiqsgx5n7oiyuxwezlo1ss6latbp0.

Schleifstein, M. (2021). $3 billion Morganza to the Gulf levee to get first $12.5 million from federal government. *nola.com, 2021*. https://www.nola.com/news/business/article_f93d27c8-5a78-11eb-9f3c-d323db4e7635.html. Accessed: 19 May 2022.

Schlottmann, A., & Miggelbrink, J. (2009). Visual geographies—an editorial. *Social Geography, 4,* (*1,* 1–11). http://www.soc-geogr.net/4/1/2009/sg-4-1-2009.pdf. Accessed: 5 January 2017.

Schlottmann, A., & Miggelbrink, J. (2015). Ausgangspunkte. Das Visuelle in der Geographie und ihrer Vermittlung. In A. Schlottmann, & J. Miggelbrink (Eds.), *Visuelle Geographien. Zur Produktion, Aneignung und Vermittlung von RaumBildern* (pp. 13–25). Bielefeld: transcript.

Schmitz, H. (1980). *Neue Phänomenologie*. Bonn: Bouvier.

Schneider, G. (1989). *Die Liebe zur Macht. Über die Reproduktion der Enteignung in der Landschaftspflege*. Kassel: Arbeitsgemeinschaft Freiraum und Vegetation.

Schneider, S. (2019). *Lehrbuch der Allgemeinen Geographie. Band 11: Luftbild und Luftbildinterpretation* (Lehrbuch der allgemeinen Geographie, Band 11, Reprint 2019). Berlin: de Gruyter.

Schneider-Sliwa, R. (2005). *USA. Geographie, Geschichte, Wirtschaft, Politik* (Wissenschaftliche Länderkunden). Darmstadt: WBG.

Schründer-Lenzen, A. (2013). Triangulation—ein Konzept zur Qualitätssicherung von Forschung. In B. Friebertshäuser, A. Langer, & A. Prengel (Eds.), *Handbuch. Qualitative Forschungsmethoden in der Erziehungswissenschaft* (pp. 149–158). Weinheim: Beltz Juventa-Verlag.

Scott, J. W., & Sohn, C. (2018). Place-making and the bordering of urban space: Interpreting the emergence of new neighbourhoods in Berlin and Budapest. *European Urban and Regional Studies, 1, (4,* 1–17). https://doi.org/10.1177/0969776418764577.

Shaffer, G. P. (2016). Decline of the Maurepas Swamp, Pontchartrain Basin, Louisiana, and Approaches to Restoration. *Water, 8, (3,* 101).

Shklar, J. N. (2020). Der Liberalismus der Furcht. In H. Bajohr (Ed.), *Judith N. Shklar. Der Liberalismus der Furcht* (3rd ed., pp. 26–66). Berlin: Matthes & Seitz.

Silliman, B. R., van de Koppel, J., McCoy, M. W., Diller, J., Kasozi, G. N., Earl, K., et al. (2012). Degradation and resilience in Louisiana salt marshes after the BP-*Deepwater Horizon* oil spill. *PNAS—Proceedings of the National Academy of Sciences of the United States of America, 109, (28,* 11234–11239).

Smith, P. (2010). The contemporary dérive. A partial review of issues concerning the contemporary practice of psychogeography. *Cultural Geographies, (17(1),* 103–122).

Sofsky, W. (2013). *Das »eigentliche Element«. Über das Böse* (Kursbuch 176: Ist Moral gut?). Hamburg: Murmann.

Soja, E. W. (1993). Los Angeles, eine nach außen gekehrte Stadt: Die Entwicklung der postmodernen Metropole in den USA. In V. Kreibich, B. Krella, U. v. Petz, & P. Potz (Eds.), *Rom—Madrid—Athen. Die neue Rolle der städtischen Peripherie* (Dortmunder Beiträge zur Raumplanung, vol. 62, pp. 213–228). Dortmund: Selbstverlag.

Solet, K. (2006). *Thirty years of change: How subdivisions on stilts have altered a Southeast Louisiana Parish's coast, landscape and people (dissertation).* (University of New Orleans Theses and Dissertations). New Orleans.

statista. (2019). Anteil der Suchmaschinen Google, Bing und Yahoo am Suchmaschinenmarkt in den USA von August 2015 bis Januar 2019. https://de.statista.com/statistik/daten/studie/152212/umfrage/anteile-von-google-bing-und-yahoo-am-us-suchmaschinenmarkt/. Accessed: 6 February 2019.

Stegemann, B. (2018). *Die Moralfalle. Für eine Befreiung linker Politik.* Berlin: Matthes & Seitz.

Steiner, C. (2009). Materie oder Geist? Überlegungen zur Überwindung dualistischer Erkenntniskonzepte aus der Perspektive einer Pragmatischen Geographie. *Berichte zur deutschen Landeskunde, 83, (2,* 129–142).

Steiner, C. (2014). *Pragmatismus—Umwelt—Raum. Potenziale des Pragmatismus für eine transdisziplinäre Geographie der Mitwelt* (Erdkundliches Wissen, vol. 155). Stuttgart: Franz Steiner Verlag.

Steinberg, T. (2006). *Acts of God. The Unnatural History of Natural Disaster in America.* Second Edition. Oxford; New York: Oxford University Press.

Stemmer, B., Philipper, S., Moczek, N., & Röttger, J. (2019). Die Sicht von Landschaftsexperten und Laien auf ausgewählte Kulturlandschaften in Deutschland—Entwicklung eines Antizipativ-Iterativen Geo-Indikatoren-Landschaftspräferenzmodells (AIGILaP). In K. Berr, & C. Jenal (Eds.), *Landschaftskonflikte* (pp. 507–534). Wiesbaden: Springer VS.

Stemmer, B., Bernstein, F., Kaußen, L., & Moczek, N. (2020). Expertenurteil und öffentliche Mitwirkung in der Landschaftsplanung und -forschung. In R. Duttmann, O. Kühne, & F. Weber (Eds.), *Landschaft als Prozess* (pp. 199–222). Wiesbaden: Springer VS.

Stemmer, B., Bernstein, F., Behre, E., & Kaußen, L. (2022). Naherholung als Teil der grünen Infrastruktur—ein neopragmatischer Ansatz. In O. Kühne, T. Freytag, T. Sedelmeier, & C. Jenal (Eds.), *Landschaft und Tourismus* (RaumFragen, in diesem Band). Wiesbaden: Springer.

Stotten, R. (2013). Kulturlandschaft gemeinsam verstehen—Praktische Beispiele der Landschaftssozialisation aus dem Schweizer Alpenraum. *Geographica Helvetica, 68, (2,* 117–127). https://doi.org/10.5194/gh-68-117-2013.

Strasser, H., & Nollmann, G. (2010). Ralf Dahrendorf. Grenzgänger zwischen Wissenschaft und Politik. *Soziologie heute, 3,* (*11,* 32–35).

Sullivan, D. (2016). Google now handles at least 2 trillion searches per year. https://searchengineland.com/google-now-handles-2-999-trillion-searches-per-year-250247. Accessed: 20 May 2019.

Swenson, D. (2021). These six factors explain why Louisiana is rapidly losing land. *nola.com.* https://www.nola.com/news/article_59675b8c-bfbe-11eb-9602-47cf4c0429dc.html. Accessed: 2 December 2021.

Templet, P. H., & Meyer-Arendt, K. J. (1988). Louisiana Wetland Loss: A Regional Water Management Approach to the Problem. *Environmental Management, 12,* (*2,* 181–192).

Theriot, J. P. (2014). *American Energy, Imperiled Coast. Oil and Gas Development in Louisiana's Wetlands.* Baton Rouge: Louisiana State University Press.

Thielmann, T. (2007). "You have reached your destination!" Position, positioning and superpositioning of space through car navigation systems. *Social Geography, 2,* (63–75).

Tilley, C. Y. (2008). *Body and image. Explorations in landscape phenomenology 2* (Explorations in Landscape Phenomenology 2). Walnut Creek, Calif: Left Coast Press.

Topper, K. (1995). Richard Rorty, Liberalism and the Politics of Redescription. *American Political Science Review, 89,* (*4,* 954–965). https://doi.org/10.2307/2082520.

Town of Grand Isle. (2018). History of Grand Isle. https://www.townofgrandisle.com/history/. Accessed: 8 July 2022.

Trepanier, J. C., & Scheitlin, K. N. (2014). Hurricane wind risk in Louisiana. *Natural Hazards, 70,* (*2,* 1181–1195). https://doi.org/10.1007/s11069-013-0869-6.

Tuan, Y.-F. (1989). *Space and Place. The Perspective of Experience* (5th ed.). Minneapolis: University of Minnesota Press.

Tuma, R. (2018). Videoanalyse und Videographie. In A. Geimer, C. Heinze, & R. Winter (Eds.), *Handbuch Filmsoziologie* (pp. 1–18). Wiesbaden: Springer VS.

Turner, R. E., Baustian, J. J., Swenson, E. M., & Spicer, J. S. (2006). Wetland sedimentation from hurricanes Katrina and Rita. *Science (New York, N.Y.), 314,* (*5798,* 449–452). https://doi.org/10.1126/science.1129116.

U.S. Geological Survey. (2022). The National Map. https://apps.nationalmap.gov/viewer/. Accessed: 4 July 2022.

Urbain, J.-D. (2003). *At the beach.* Minneapolis: University of Minnesota Press.

Urry, J., & Larsen, J. (2011). *The Tourist Gaze 3.0.* Los Angeles: SAGE Publications.

van Essen, F. (2013). *Soziale Ungleichheit, Bildung und Habitus.* Wiesbaden: Springer VS.

Vester, H.-G. (1993). *Soziologie der Postmoderne.* München: Quintessenz.

Wagner, E. (2019). *Intimisierte Öffentlichkeiten. Pöbeleien, Shitstorms und Emotionen auf Facebook.* Bielefeld: transcript.

Waldenfels, B. (2000). *Das leibliche Selbst. Vorlesungen zur Phänomenologie des Leibes* (Suhrkamp-Taschenbuch Wissenschaft, vol. 1472). Frankfurt (Main): Suhrkamp.

Walz, U. (2001). *Charakterisierung der Landschaftsstruktur mit Methoden der Satelliten-Fernerkundung und der Geoinformatik.* Baden-Baden: Nomos.

Wardenga, U. (2001a). Theorie und Praxis der länderkundlichen Forschung und Darstellung in Deutschland. In F.-D. Grimm, & U. Wardenga (Eds.), *Zur Entwicklung des länderkundlichen Ansatzes* (Beiträge zur Regionalen Geographie, vol. 53, pp. 9–35). Leipzig: Selbstverlag.

Wardenga, U. (2001b). Zur Konstruktion von 'Raum' und 'Politik' in der Geographie des 20. Jahrhunderts. In P. Reuber, & G. Wolkersdorfer (Eds.), *Politische Geographie. Handlungsorientierte Ansätze und Critical Geopolitics* (Heidelberger geographische Arbeiten, vol. 112, pp. 17–32). Heidelberg: Selbstverlag des Geographischen Instituts der Universität Heidelberg.

Wardenga, U. (2002). Alte und neue Raumkonzepte für den Geographieunterricht. *Geographie heute, 23,* (*200,* 8–11).

Wardenga, U. (2006). German Geographical Thought and the Development of *Länderkunde. Inforgeo, 18,* (127–147).

Warms, C. A., & Schroeder, C. A. (1999). Bridging the gulf between science and action: The "new fuzzies" of neopragmatism. *Advances in Nursing Science, 22,* (*2,* 1–10).

Warnke, M. (1992). *Politische Landschaft. Zur Kunstgeschichte der Natur.* München: Carl Hanser Verlag.

Wartmann, F. M., & Mackaness, W. A. (2020). Describing and mapping where people experience tranquillity. An exploration based on interviews and Flickr photographs. *Landscape Research, 45,* (*5,* 662–681). https://doi.org/10.1080/01426397.2020.1749250.

Weber, M. (1972 [1922]. *Wirtschaft und Gesellschaft. Grundriss der verstehenden Soziologie* (5., revidierte Auflage). Tübingen: J.C.B. Mohr (Paul Siebeck).

Weber, M. (1976 [1922]. *Wirtschaft und Gesellschaft. Grundriß der verstehenden Soziologie.* Tübingen: Mohr Siebeck.

Weber, M. (2010 [1904/05]. *Die protestantische Ethik und der Geist des Kapitalismus* (Beck'sche Reihe, vol. 1614, Vollständige Ausgabe, 3. Auflage). München: C.H. Beck.

Weber, F., Jenal, C., Roßmeier, A., & Kühne, O. (2017). Conflicts around Germany's *Energiewende*: Discourse patterns of citizens' initiatives. *Quaestiones Geographicae, 36,* (*4,* 117–130). https://doi.org/10.1515/quageo-2017-0040.

Weber, F., Kühne, O., Jenal, C., Aschenbrand, E., & Artuković, A. (2018). *Sand im Getriebe. Aushandlungsprozesse um die Gewinnung mineralischer Rohstoffe aus konflikttheoretischer Perspektive nach Ralf Dahrendorf.* Wiesbaden: Springer VS.

Weichhart, P. (1999). Die Räume zwischen den Welten und die Welt der Räume. In P. Meusburger (Ed.), *Handlungszentrierte Sozialgeographie. Benno Werlens Entwurf in kritischer Diskussion* (Erdkundliches Wissen, vol. 130, pp. 67–94). Stuttgart: Steiner.

Weingart, P. (2015). *Wissenschaftssoziologie.* Bielefeld: transcript.

Weißmann, M., Edler, D., & Rienow, A. (2022). Potentials of Low-Budget Microdrones: Processing 3D Point Clouds and Images for Representing Post-Industrial Landmarks in Immersive Virtual Environments. *Frontiers in Robotics and AI,* (*9,* 1–12). https://doi.org/10.3389/frobt.2022.886240.

Welsch, W. (2002). *Unsere postmoderne Moderne* (6. Auflage). Berlin: Akademie-Verlag.

Wendland, T. (2018). Lack Of Funds Keeps Louisiana From Buying Out Coastal Residents. *WWNO—New Orleans Public Radio.* https://www.npr.org/2018/01/05/575876626/lack-of-funds-keeps-louisiana-from-buying-out-coastal-residents?t=1659460233149. Accessed: 2 August 2022.

Werlen, B. (1986). Thesen zur handlungstheoretischen Neuorientierung sozialgeographischer Forschung. *Geographica Helvetica, 41,* (*2,* 67–76). https://doi.org/10.5194/gh-41-67-1986.

Werlen, B. (1997). *Gesellschaft, Handlung und Raum. Grundlagen handlungstheoretischer Sozialgeographie* (Erdkundliches Wissen, vol. 89). Stuttgart: Steiner.

Werlen, B. (2008). Körper, Raum und mediale Repräsentation. In J. Döring, & T. Thielmann (Eds.), *Spatial Turn. Das Raumparadigma in den Kultur- und Sozialwissenschaften* (pp. 165–192). Bielefeld: transcript.

Winchester, H. P. M., Kong, L., & Dunn, K. (2003). *Landscapes. Ways of imagining the world.* London: Routledge.

Wittgenstein, L. (1995 [1953]). *Tractatus logico-philosophicus. Tagebücher 1914–1916. Philosophische Untersuchungen* (10. Auflage). Frankfurt (Main): Suhrkamp (Werkausgabe Band 1).

Wojtkiewicz, W. (2015). *Sinn—Bild—Landschaft. Landschaftsverständnisse in der Landschaftsplanung: eine Untersuchung von Idealvorstellungen und Bedeutungszuweisungen.* Berlin: Technische Universität Berlin.

Wojtkiewicz, W., & Heiland, S. (2012). Landschaftsverständnisse in der Landschaftsplanung. Eine semantische Analyse der Verwendung des Wortes "Landschaft" in kommunalen Landschaftsplänen. *Raumforschung und Raumordnung—Spatial Research and Planning, 70,* (*2,* 133–145). https://doi.org/10.1007/s13147-011-0138-7.

Wood, D., & Fels, J. (1986). Design on Signs/Myth and Meaning in Maps. *Cartographica: The International Journal for Geographic Information and Geovisualization, 23,* (*3,* 54–103). https://doi.org/10.3138/R831-50R3-7247-2124.

Wunderlich, W. E. (1964). Water hyacinth control in Louisiana. *Hyacinth Control Journal / Journal of Aquatic Plant Management, 3,* (4–7).

WXChasing. (2021). Grand Isle, La Drone video of Hurricane Ida Damage whole Island- Category 4 4k. https://www.youtube.com/watch?v=4CYPuzlotpk. Accessed: 30 June 2022.

Wylie, J. (2005). A single day's walking: narrating self and landscape on the South West Coast Path. *Transactions of the Institute of British Geographers, 30,* (*2,* 234–247). https://doi.org/10.1111/j.1475-5661.2005.00163.x.

Wylie, J. (2007). *Landscape.* Abingdon: Routledge.

Yeo, J., & Knox, C. C. (2019). Public Attention to a Local Disaster Versus Competing Focusing Events: Google Trends Analysis Following the 2016 Louisiana Flood. *Social Science Quarterly, 100,* (*7,* 2542–2554). https://doi.org/10.1111/ssqu.12666.

Yodis, E. G., Colten, C. E., & Hemmerling, S. E. (2016). *Geography of Louisiana* (7. Auflage). Boston: McGraw-Hill.

Zepp, H. (2020). Das Neue Emschertal. Transformation von Freiräumen und Veränderung von Ökosystemleistungen während der letzten 200 Jahre. In R. Duttmann, O. Kühne, & F. Weber (Eds.), *Landschaft als Prozess* (pp. 327–360). Wiesbaden: Springer VS.

Ziemann, A. (2012). *Soziologie der Medien* (2., überarbeitete und erweiterte Auflage). Bielefeld: transcript Verlag.

Zierhofer, W. (1999). Geographie der Hybriden. *Erdkunde, 53,* (*1,* 1–13).

Zierhofer, W. (2002). *Gesellschaft. Transformation eines Problems* (Wahrnehmungsgeographische Studien, vol. 20). Oldenburg: Bibliotheks- und Informationssystem der Univ.

Zierhofer, W. (2003). Natur—das Andere der Kultur? Konturen einer nicht-essentialistischen Geographie. In H. Gebhardt, P. Reuber, & G. Wolkersdorfer (Eds.), *Kulturgeographie. Aktuelle Ansätze und Entwicklungen* (Spektrum Lehrbuch, pp. 193–212). Heidelberg: Spektrum Akademischer Verlag.

Zimmermann, S. (2007). Media Geographies: Always Part of the Game. *Aether—the journal of media geography, 1,* (59–62).

Zimmermann, S. (2019). Filmlandschaft. In O. Kühne, F. Weber, K. Berr, & C. Jenal (Eds.), *Handbuch Landschaft* (pp. 623–629). Wiesbaden: Springer VS.

Zoglauer, T. (1998). *Geist und Gehirn. Das Leib-Seele-Problem in der aktuellen Diskussion* (UTB für Wissenschaft. Uni-Taschenbücher: Philosophie, vol. 2066). Göttingen: Vandenhoeck & Ruprecht.

Zonn, L. (Ed.). (1990). *Place Images in Media. Portrayal, Experience, and Meaning.* Savage, Maryland: Rowman & Littlefield Publishers, inc.

Printed in the United States
by Baker & Taylor Publisher Services